TECHNOLOGY VENTURING

OTHER BOOKS IN THE SERIES:

Commercializing Defense-Related Technology, edited by Robert L. Kuhn

Corporated Creativity: Robust Companies and the Entrepreneurial Spirit, edited by Raymond W. Smilor and Robert L. Kuhn

TECHNOLOGY VENTURING: AMERICAN INNOVATION AND RISK-TAKING

edited by
Eugene B. Konecci
and Robert Lawrence Kuhn

PRAEGER

PRAEGER SPECIAL STUDIES • PRAEGER SCIENTIFIC

New York • Philadelphia • Eastbourne, UK
Toronto • Hong Kong • Tokyo • Sydney

Library of Congress Cataloging-in-Publication Data
Main entry under title:

Technology venturing.

Derived from a conference held at the University of Texas at Dallas, Feb. 5–7, 1984.
Includes index.
1. Technological innovations—United States—Congresses. 2. Venture capital—United States—Congresses. I. Konecci, Eugene B. II. Kuhn, Robert Lawrence.
HC110.T4T45 1985 338'.06 85-12192
ISBN 0-03-005183-5 (alk. paper)

Published in 1985 by Praeger Publishers
CBS Educational and Professional Publishing, a Division of CBS Inc.
521 Fifth Avenue, New York, NY 10175 USA

56789 052 987654321

Printed in the United States of America on acid-free paper

INTERNATIONAL OFFICES

Orders from outside the United States should be sent to the appropriate address listed below. Orders from areas not listed below should be placed through CBS International Publishing, 383 Madison Ave., New York, NY 10175 USA

Australia, New Zealand
Holt Saunders, Pty, Ltd., 9 Waltham St., Artarmon, N.S.W. 2064, Sydney, Australia

Canada
Holt, Rinehart & Winston of Canada, 55 Horner Ave., Toronto, Ontario, Canada M8Z 4X6

Europe, the Middle East, & Africa
Holt Saunders, Ltd., 1 St. Anne's Road, Eastbourne, East Sussex, England BN21 3UN

Japan
Holt Saunders, Ltd., Ichibancho Central Building, 22-1 Ichibancho, 3rd Floor, Chiyodaku, Tokyo, Japan

Hong Kong, Southeast Asia
Holt Saunders Asia, Ltd., 10 Fl, Intercontinental Plaza, 94 Granville Road, Tsim Sha Tsui East, Kowloon, Hong Kong

Manuscript submissions should be sent to the Editorial Director, Praeger Publishers, 521 Fifth Avenue, New York, NY 10175 USA

Preface

"Technology venturing" is a fresh concept; it integrates applied science and risk capital in the development of large-scale, long-term projects of national importance. It is an entrepreneurial process by which major institutions take and share risk in combining and commercializing scientific research and technological innovation. It is a primary means of generating original products and services of economic value, often linking public-sector requirements with private-sector investments within a collaborative competitive environment.

What entrepreneurship can do for small business, technology venturing can do for corporations, universities, consortiums, and government at all levels. Thus technology venturing is a catalyst for a broad-based, emergent entrepreneurial spirit in America. Traditional methods of financial analysis in the private sector or cost-benefit analysis in the public sector are no longer appropriate in this new domain. Technology venturing demands individual initiative and collective commitment, and much of America's economic hope and social sustenance is riding right with it.

This volume is derived from a conference of the same name ("Technology Venturing: American Innovation and Risk-Taking") held at the University of Texas at Dallas, February 5–7, 1984, and sponsored by the IC^2 Institute of the University of Texas at Austin and the RGK Foundation.

The overall mission of the IC^2 Institute is to strengthen the American economy through three primary factors: innovation, creativity, and capital. We seek national consensus across diverse sectors for establishing and maintaining America's economic, technological, and scientific leadership in the world. The previous volume in this series, *Commercializing Defense-Related Technology* (Praeger, 1984), stressed the importance of economic elements in building comprehensive national security and highlighted the need for intersector cooperation among industry, university, and government.

The current volume examines the changing basis for the U.S. economy, viewing technology venturing as the leading edge of American dynamism. Two themes form our focus: First, there is an American response for improving our domestic economy and meeting international competition, including maintaining scientific/technological preeminence. To compete effectively we need to find creative and innovative ways to unify public-sector initiatives with private-sector incentives through technology venturing.

Second, there is a need for coherent policies and initiatives that foster new jobs, stimulate economic growth, encourage emerging industries, renew basic industries, and forge stronger links among academia, business, and government. These policies should encourage innovative roles for the public and private sectors in the formulation and implementation of large-scale programs and in the development of small- and medium-sized businesses. In this context, several papers highlight forthcoming large-scale programs concerned with national security, space commercialization, and rebuilding the public infrastructure.

As an entrepreneurial process, technology venturing commercializes science and technology through newer institutional arrangements. In an environment in which the public sector serves as stimulator for technological development through novel state and local policies and initiatives, technology venturing can provide opportunities for the private sector to take the leadership role in a changing economy.

Six parts and an appendix comprise the book: (I) The Changing Economy (the macroeconomic setting and need for collaboration); (II) Initiatives for Quantum Changes (policy recommendations for federal, and state legislation, and private-sector leadership); (III) National Policies (challenges and strategic implications); (IV) State Policies (Initiative development); (V) Large Scale Programs (National and International problems and opportunities in public works and space); and (VI) The Next Steps in Large Scale Programs (private/public collaborative efforts); the Appendix contains the Data Book (information on research and development program, Large-Scale projects and cooperative technology venturing efforts).

We at the IC^2 Institute dedicate this work to the staffs and officials of federal, state, and local government agencies who are responsible for coordinating the public-sector side of the equation. We also appreciate the participation of our contributors, many from the private sector, whose ideas we believe to be vital. We pledge to continue catalyzing the cooperative interaction among all sectors of America as we work together to build a stronger economy and better society.

Eugene B. Konecci
Robert Lawrence Kuhn

Austin, Texas
March 1985

Acknowledgments

Technology venturing is an integrative risk-taking and risk-sharing approach to the commercialization of scientific knowledge and technology. Many individuals, institutions, and groups are responsible for the success of the conference that led to this book.

Credit must be given to Dr. George Kozmetsky and Michael Gill for coining the term "technology venturing" and promoting the future potential and expansion of a fresh and dynamic approach to the process. Dr. Kozmetsky has devoted his energies and resources to the encouragement and support of public- and private-sector researchers, leaders, and institutions to cooperate and collaborate in the updating and creating of newer integrated bodies of knowledge and state-of-the-art technologies.

This book is the result of the devotion and efforts of many individuals and organizations in addition to the conference participants. Foremost, we express our thanks and gratitude to Ronya Kozmetsky, president of the RGK Foundation, for her outstanding skills and abilities in arranging and handling the logistics of such a complex endeavor.

The planning committee worked closely with Dr. Kozmetsky and Dr. Konecci in providing the inputs and critiques so needed in making the conference such a success. The committee included the following: Dr. Richard D. DeLauer, under secretary of defense for research and development, U.S. Department of Defense; Maxwell Hunter, II, Lockheed; Rolf Illsley, chairman and CEO, Optical Coating Laboratories; Dr. Edward Kelly, NSF; Dr. J.R. Kirkland, F.B.A., Inc.; Dr. Raymond Smilor, associate director of the IC^2 Institute, University of Texas at Austin; and Dr. Eugene Stark, chairman, Federal Laboratory Consortium, Los Alamos National Laboratory.

Special thanks must go to all the participants of the conference. They are listed in the order of their participation: A. Starke Taylor, Mayor of Dallas; Jon Newton, chairman of the Board of Regents, University of Texas System; The Honorable Mark White, governor of Texas; Joe Dealy, Sr., chairman of the board, *Dallas Morning News*; Robert Ortner, chief economist, U.S. Department of Commerce; Cordell Hull, vice-president, and chief financial officer, Bechtel Corporation; Francis Cotter, vice-president, Government Affairs, Westinghouse Electric Corporation; The Honorable Lloyd Bentsen, U.S. senator (Texas); William Gregory, editor-in-chief, *Aviation Week and Space Technology*; Dr. James Wade, Department of Defense; John Egan, Coopers and Lybrand; Doug Pitcock, president, Williams Brothers Construction; D. Keith Dodson, vice-president Brown & Root; General Robert J. Dacey, U.S. Army Corps of Engineers; Henry G. Cisneros, mayor of San Antonio; Dr. Eugene B. Konecci, The Kleberg-King Ranch Professor, University of Texas at Austin; Dr. Eugene Stark, Los Alamos National Laboratory; Rolf Illsley, chairman, Optical Coatings Laboratories Inc.; Emil Sunley, Deloitte Haskins & Sells; Dr. Ken Gordon, President's Commission on Industrial Competition; Dr. Frank Davidson, professor, Macroengineering Research Group, MIT; Dr. George Kozmetsky, director the IC^2 Institute, The University of Texas at Austin; Admiral Bobby Ray Inman (Retired) chairman, CEO, and president Microelectronics and Computer Technology Corporation; Dr. E. Donald Walker, chancellor, the University of Texas System; Dr. Victor Arnold, Bureau of Business Research, the University of Texas; Dr. Harden Wiedemann, director, Economic Development, state of Texas; Dr. Wendell Nedderman, president, University of Texas at Arlington; Dr. Duane Leach, president, University of Texas of the Permian Basin; Captain Rick Cantwell, U.S. Army Corps of Engineers; David T. Thompson, Deloitte Haskins & Sells; Joseph Kigen, Westinghouse Electric Corporation; Dr. Robert Kuhn, senior research fellow, the IC^2 Institute.

Contributors of the Technology Venturing Data Book included Dr. Kozmetsky, Dr. Konecci, Dr. Smilor, Michael Gill, Kathleen J. Murphy, Dr. Victor Arnold, Dr. Hardin Wiedemann, Malcolm Sherman, Dr. Duke Perreira, Dr. W. Arthur Porter, and Edward O. Vetter.

We appreciate the assistance of the RGK Foundation and the IC^2 Institute staffs for their devotion to the many tasks that needed to be accomplished in support of the meeting. These included Michael Gill, Cynthia Smith, Ophelia Mallari, Myrna Braziel, Diane Kane, Patricia Roe, Susan Pressler, Christie Nelson, Margo Latimer, and Cathrine Alleman. Special thanks go to David Smith for his technical and electronic abilities in handling the audio-visual aspects and the recording of the conference.

Finally, we wish to express our thanks to George Zimmer, editor at Praeger, for his constructive suggestions and insights in the development of this book.

E.B.K.
R.L.K.

Contents

Part V
Large-Scale Programs: Problems and Opportunities

Part VI
Large-Scale Programs: The Next Steps

PART I

THE CHANGING ECONOMY: THE NEED FOR COLLABORATIVE COMPETITION

1

Technology Venturing: The New American Response to the Changing Economy

George Kozmetsky

This volume on technology venturing sets forth a new American response to our changing economy. This response is a grassroots approach for strengthening U.S. leadership in a global marketplace.

Technology venturing has come about because of a series of key stimulants. These include:

1. Federal government programs stimulating risk-taking investments.
2. State government initiatives fostering high-technology growth.
3. Imaginative collaborative relationships between universities and corporations.
4. Pioneering programs and linkages among government, business, and universities.
5. Innovative private joint efforts for scientific advances.
6. A blossoming venture capital industry.
7. Creative institutional arrangements among the public, private, and nonprofit sectors.
8. Fresh local community concepts for economic growth and social developments.
9. Worldwide technology competition and the struggle for scientific preeminence.
10. Expanding needs for large-scale programs on a global basis.

These stimulants are also creating a new economy — an economy that emphasized sustained productivity over rapid expansion, adaptability and flexibility over efficiency and effectiveness, and quality and choice over quantity and rigidity. It is an economy in which growth is based on technology diversification and the needs of a demographically shifting population. Competition within global markets is economic, scientific, and technological in nature. Achieving competitive advantage in the new economy is creating the new American innovation — technology venturing.

Technology venturing is the modern mechanism for forming new jobs and promoting economic growth. It is also a practical way to deal with the besetting problems being discussed in the media, halls of Congress, state legislatures, and in the private sector. It is based on the unique, American entrepreneurial spirit that has always depended upon experience, risk taking, and common sense. The scientific, technological, and economic preeminence of the United States depends on the strength, vigor, and dynamism of our own unique culture and heritage. The United States has maintained, and will continue to maintain, its leadership because of our abilities to encourage the emergence of technology-based businesses, expand venture capital markets, restructure and reform financial and security markets, and collaborate as required between public- and private-sector institutions. America's strength has always been an ability to be scientifically creative, technologically adept, managerially innovative, and entrepreneurially daring. In these ways, we have met and will continue to meet critical challenges to promote the common good and support the general welfare.

Do not misconstrue technology venturing. It is not conservative versus liberal; not capitalism versus socialism; not automated machines versus the human use of human beings. It is not technocrats versus nontechnocrats; not scientific elite versus blue collar; not a service economy versus an industrial economy.

Rather, technology venturing is a new American response to the changing worldwide economy. It is both industrial renewal and a revival of American spirit. It emphasizes coherent and collaborative strategies that can generate an economic renaissance for the next American frontier.

Technology venturing, when combined with innovative approaches to investing in the public infrastructure, space commerce, and other large-scale projects, provides a balanced and dynamic integrative framework for long-range human resource development, sustained economic growth, and job security.

Technology venturing is an integrative process. It incorporates a dynamic private sector, a creative role for government through federal, state, and local initiatives and programs, and newer academic relationships. It also fosters corporate and community collaborative efforts while nurturing positive government/academic/business relationships.

Technology venturing helps improve our educational structure, fulfill critical manpower requirements, and enhance our industrial creativity and innovation. It is primarily a means for encouraging the emergence of a myriad of leading edge businesses in the context of a private enterprise system that has always been our own way to achieve and maintain U.S. preeminence.

The remainder of this chapter examines the role and scope of technology venturing from three perspectives:

1. The strategic implications of a changing economy.
2. Emerging large-scale programs.
3. Innovative, collaborative, and institutional relationships.

As a result of this examination, we can begin to lay out a possible road map for technology venturing — directions necessary if it is truly to meet its promise for American innovation, risk taking, and risk sharing. Such a road map offers multiple perspectives.

STRATEGIC IMPLICATIONS

The following issues are central to understanding the strategic implications of a changing economy:

1. We will continue to live under high federal deficits for a while.
2. The changing economy will require extensive and interactive capital investment commitments by each sector of our economy. Each sector must understand and appreciate the needs of the others and then be individually willing to meet those needs.
3. Cohesive national and state policies are needed over a longer term, certainly on the order of five to ten years.
4. We will need to encourage more entrepreneurship in individual start-ups and more intrapreneurship in established companies if we as a nation are to capitalize on the myriad of opportunies generated by technology venturing.

High federal deficits should not deter our nation's resolve to meet pressing demands. We must:

1. Provide for adequate national security.
2. Stimulate the rebuilding of American industry.

3. Ensure adequate employment for all segments of our population in a reasonable time.
4. Enhance the growth of new enterprises.
5. Improve and advance the educational system.
6. Strengthen the U.S. position in global markets.
7. Assure productive competition among the participants in technology venturing.

These are a tough set of goals. However, we have no alternative but to achieve them.

Fortunately, the current success of technology venturing indicates that it is possible to accomplish these stringent goals. Our changing economy requires longer-term capital commitments. Put another way, the old American economy is becoming financially and technologically obsolete. We are seeing a vigorous capital infusion into the construction of a newer economic infrastructure. It is not surprising that this involves federal and state governments, businesses, and universities all making investments simultaneously. Piecemeal will not work. We cannot develop a new economic infrastructure by letting one sector change at a time or at the expense of the other. We need to move on a broad front of capital investments. Coherence is key.

A study by the Associated General Contractors of America recommended over $1 trillion of cumulative costs for repair, replacement, and renewal of the U.S. infrastructure for the next 19 years. The Reagan budget for fiscal year 1985 as submitted recommended $153 billion for the next five years. If the five-year average is used, it will take about 50 years to build a minimum public infrastructure. Such a time frame will not be acceptable to the American public.

The current NASA and Department of Defense space and military programs also are longer-term investments. State commitments to change education are likewise long range. These are all multibillion dollar commitments. In addition, private-sector initiatives, especially those involving macroprojects, demand long-term thinking. Individually, many of them will call for at least $100 million to over $10 billion. In other words, if we are to have a viable new economy in the 1990s and beyond, many macroprojects will have to be undertaken to support the emerging industrial, public, and educational infrastructures.

As part of our educational requirements to achieve scientific preeminence, for example, we must eliminate technological illiteracy and extend scientific, engineering, and management capabilities. Only through an enhanced educational system can we maintain a comparative advantage over other nations in the world marketplace. These commitments require time to develop excellent faculties at all educational levels and responsive curricula and courses.

The changing economy strategically requires us to reexamine, if not modify and in some cases drop, traditional standards for judging appropriateness of national and state budget policies as well as public and private investment decisions. Neither deficits alone nor noneconomic decisions cause living beyond our means or potential bankruptcy. What matters most is whether we have to provide the investment in terms of priorities for today and tomorrow. If deficits are used to develop a required public infrastructure that provides modern, safe, and economicial transportation and water systems, they can well pay for themselves over time through both savings of current expenditures as well as broadening our economic base.

When one asks if an investment is economical, we can't simply focus on reviewing its return on investment or assessing the present value of its income stream in relation to the risk undertaken. This conceptual attitude is crucial. In an attempt to be analytical and quantitative, we can miss the point and cloud the goal.

Just as important as the usual economic standards of effectiveness and efficiency are two other dimensions. First, *flexibility*. If we don't make the investments, will we still have sufficient flexibility to meet competitive challenges foreign or domestic? Second, *adaptability*. If we don't make these investments, can we adapt ourselves to unexpected or unforeseen competition without serious consequences to our firms and communities? These are the questions that the new economy raises. Technology venturing permits us to attack problems in a collaborative way.

As Dr. Albert M. Wojnilower, managing director of the First Boston Corporation, has stated:

> If trees are to be planted whose shade is to be enjoyed by our heirs, we need to choose the right trees (whether or not they happen to be labeled "investment"); to find a mutually caring and respectful balance between young and old; and to avoid undue exploitation by or of others. Whether progress toward these goals increases or reduces the budget deficit is immaterial.[1]

---

[1]Wojnilower, Albert J., "Don't Blame the Deficit." Presented at the October 1983 Bald Peak Conference sponsored by the Federal Reserve Bank of Boston.

EMERGING LARGE-SCALE PROGRAMS

This topic of macroengineering projects (or "macro") has not, for some reason, had an adequate public forum during the past decade. Yet management of large complex projects has been — and will continue to be — a unique American resource, even while the gap is being closed by Japan, France, West Germany, and other nations.

There are four major reasons to focus on large-scale macroprojects in terms of the new economy:

1. They could have a direct and immediate impact on job creation that is not just "make work." Furthermore, they could provide employment for all segments of the population requiring a range of skills from low specialized manual to highly specialized scientific and technical tasks. This employment is productive and can produce immediate retraining on the job that can later be transformed to other opportunities.
2. They are required to strengthen the national security if we are to reduce the possibility of international conflict by improving our space defense capabilities.
3. They enhance the commercialization of science and technology. Macroprojects can produce thousands of micropossibilities. They can also provide for a coherent network of multiscientific inventions to stimulate improvements as well as new technology developments for basic and emerging industries.
4. There is a viable and large international market for macroengineering projects. In the 1970s there were over 1,615 projects worldwide with a planned investment of over $1 trillion. The emerging macroproject needs for the 1980s are in energy, transportation, urban development, materials processing, and agricultural developments. From 1980 to 1983, there were over $375 billion of planned investments for Third-World macroprojects alone.

Macroprojects are often technological ventures. These large-scale efforts require innovative collaborative and coherent policies, concurrent operations, and significant investment from both public and private sources. They are enterprises that are long term in nature; most importantly, they can build technology as a resource that can then be commercialized in many directions.

An American space station program, for example, would have spinoffs in artificial intelligence, electromechanical systems (robotics and teleoperators), integrated design, simulation, and analysis tools, advanced control system architecture, submicron microelectronics, and new methods for construction.

What may be needed for successful macroprograms is to expand the financial risks to minimize the overall risk, especially if technology can be spun off. At the same time, macroprograms may allow us to utilize better in the short run our scarce scientific and engineering human resources.

Macroprojects can play an important role in helping to put our nation's economy back into equilibrium and stable growth. The key is to develop multiple uses from our scientific, technological, and human resources through the commercialization process. We need better management and more effective early planning to spin off government-initiated programs. We need better mechanisms and ways to conduct technology venturing that reduce financial and technological risk taking while expanding risk sharing across a broader base of recipients.

INNOVATIVE COLLABORATIVE EFFORTS

The third perspective from which to examine technology venturing is to review innovative collaborative relationships.

Throughout America, our states are moving to stimulate high-technology industry and to foster technological commercialization. They, rather than the federal government, have taken a leadership role through policy development organizations, economic growth initiatives, and corporate-university partnerships. Over 150 programs of initiatives have been developed by states for high-technology and economic development in the past few years. At least 15 states have instituted high-technology education programs. The major source of research and development (R&D) funds, however, is still the federal government. There is a direct correlation between those states receiving the largest federal obligations for R&D and those taking the lead to initiate high-technology development. (The only exceptions to this general rule are North Carolina, Georgia, Minnesota, and Utah. They are utilizing more of their own state funds.)

We are seeing the emergence of technopolies in most states. These technopolies are bringing together, in dynamic and interactive ways, state government, local government, private corporations, universities, and nonprofit foundations and other organizations. Crescents, corridors, and triangles are developing between key cities or research universities. We are seeing centers of excellence within these geometric areas. Local leaders have begun to lay out science and research parts; they have begun to target emerging science and technologies for long-term industrial growth and vitality. Leadership networks are forming between previously isolated institutions.

Such technopolies are not accidental. They can cause catalytic reactions far beyond their immediate location within the state. Very often these impacts are national and international in scope. It is quite an exhilarating and rewarding experience to be part of such an innovative risk-taking adventure.

Socially conscious leaders are making this happen. They are blurring the traditional boundaries between institutions. They collaborate; they share; and they are willing to take personal and institutional risks. They have found out through experience that technology as a resource truly has the attribute that the more you use it, the more you have. Technology venturing is not a zero sum game.

Technology venturing is spurring a new breed of innovators. They are tapping a burgeoning capital venture industry, R&D partnerships, and new business incubators to cultivate companies. Entrepreneurial teams are also focusing on tying together expertise in marketing, engineering, management, and finance. Academics are setting up their own firms. Capital venturers are growing $100 million companies as well as transnational organizations in the $1 billion or more range. Large firms are emphasizing intrapreneurship to meet the demands of international competition and changing personnel desires.

We are in the early stages of technology venturing. It is an emerging phenomenon. It needs to be tempered and strengthened. There are good reasons to keep this momentum going. A recent Government Accounting Office (GAO) study has shown the importance of venture capital in providing societal benefits. The study sets forth that 72 successful companies were started by an aggregate investment of $209 million. These companies provided the following aggregate economic benefits:

1. Aggregate sales in 1979 alone of $6 billion which, in the latter five years (1974–79), were growing at an annual rate of 33 percent.
2. An estimated 130,000 jobs.
3. Over $100 million in corporate tax revenues.
4. $350 million in employee tax revenues.
5. $900 million in export sales.

By 1989, the GAO estimates that venture-backed companies could generate the following amounts:

1. $3 billion to $7.6 billion in corporate taxes.
2. $8 billion to $22.7 billion in employee taxes.
3. $26 billion to $81.9 billion to export sales.
4. 522,000 to 2,244,000 jobs.

In addition, the GAO noted that these venture-backed firms brought to market primarily productivity-enhancing technology products. This fact signifies that venture capital firms have an impact not only on the stability and productivity of firms they finance but also on other segments of the economy. The extent to which venture capital investment amplifies other segments of the economy is impressive. The GAO estimates that for every $1,000 of venture capital invested in the 1970s, $40,000 to $54,000 worth of productivity-enhancing products and services will be sold during the 1980s. The GAO report conclusively demonstrates the truly dynamic nature of the venture capital investment process.

An impressive range of university-corporation partnership arrangements are evolving. A majority of these joint programs are with private universities, but public university-corporation programs are also emerging rapidly.

In response to fierce international competition, American corporations are finding innovative ways to collaborate, especially in R&D. These ties are challenging some of our nation's most fundamental notions about competition, antitrust, and technology transfer.

At the federal level, numerous types of joint ventures between governments are under consideration. They involve space programs, defense agendas, infrastructure rebuilding, import-export relationships, and the very nature of foreign policy. Consequently, we need to extend our foreign policy to include a viable *technology policy*. Such a policy would extend technology influence beyond security and defense or even technology transfer to economic, social, and political sectors. Such a policy would examine American employment in terms of technology transfer and foreign trade.

CHARTING TECHNOLOGY VENTURING

In a real sense, technology venturing is a pioneering task, much like that of early explorers who sought to discover their environment. As Daniel J. Boorstin, historian and now librarian of Congress, has stated:

> The great obstacle to discovering the shape of the earth, the continents, and the ocean was not ignorance but the illusion of knowledge. Imagination drew in broad strides, instantly serving hopes and fears, while knowledge advanced by slow increments and contradicting witnesses.[2]

Let us proceed with imagination, hope, and daring to chart the direction of technology venturing. In this task, as in earlier explorations, our map of this previously uncharted territory will leave much for the imagination.

Critical specific needs must be initiated on a national basis now. Among these are:

1. Advancement of the purposes of the American Society for Macroengineering. The society is a private nonprofit organization concerned with the critical issues related to the development of macroengineering projects, programs, and systems, such as planned cities, industrial complexes, energy projects, regional development projects, outer space programs, and other enterprises requiring capital investment ranging from hundreds of millions to billions of dollars.
2. Establishment of large-scale project research consisting of a consortium of universities to study large-scale engineering-management programs and projects.
3. Identification of a National Technology Agenda. As we adjust to the new global economy, we must identify barriers, limitations, and problems to delineate the role of technology. Some of the more specific concerns take a worldwide perspective.
4. Assurance that spin-off opportunities become major programs. This includes in the case of government programs the transfer of technology to the private sector.
5. Establishment of research projects in universities that seek better solutions to public infrastructure and water problems. This includes such special projects as new materials and techniques to reduce costs and improve the effectiveness of our maintenance programs as well as construction and operations.
6. Enhancement of private-public venturing activities and opportunities. This includes better ways of mitigating unforeseen externalities as well as risk-taking mechanisms and risk-sharing approaches.

---

[2]Borstin, Daniel J., *The Discoverers* (New York: Random House, 1983) p. 86.

7. Transformation of our educational structure for the new economy.

Not all the initiatives need or should come from Washington, D.C. There are important leadership roles for state and local governments, universities, and private companies transforming the economy that demand multifaceted initiatives.

I firmly believe that we are at a great watershed point in American history. This momentous period is fully as important as the very founding of our nation. Just as those who built this wonderful land needed to harness the strength of will, experience, daring, knowledge, and foresight to start a great nation, so we, too, must become the modern founders of a new American society robust and healthy for generations to come.

As a Texan, I am proud that we have taken a leadership role in establishing a new era in technology venturing. On the other hand, we still have much to do to strengthen technology venturing at home. We need to establish more community programs such as that in the San Antonio-Austin Corridor (Austin in computers and electronics; San Antonio in medicine and biotechnology). We need to determine how best to establish viable transportation systems especially in view of the fact that our research universities and many of the emerging technology-based companies are located in the cities currently without adequate transportation accessibility for future economic growth. We need to establish an advanced telecommunications system for educational purposes for primary, secondary, and higher education. This can do much to reduce the future tax burden and at the same time provide an important example for other states to participate and to use. We need to recognize that over 40 percent of our population live and work outside of Dallas-Fort Worth, Houston, San Antonio, Austin, and El Paso. They, too, need to be linked into the Texas technology venturing network.

2

Comprehensive National Security: The Power of American Science

Robert Lawrence Kuhn[1]

Science separates present from past. It is the critical difference between savages living like animals and humans living like people. Science is more than a subject in school; it is the foundation of our world, the progenitor of present-day society, the source of contemporary civilization — in short, science is axial to our way of life.

Science is both process and content, the mechanism of discovery as well as the thing discovered. The scientific method is the core paradigm of modern man; it is the shortest distance and surest route to factual truth, the line of thinking most logical and reproducible. The scientific method is perhaps mankind's finest conceptual tool: unbiased data collection; creative hypothesis generation (induction); rigorous analytical reasoning (deduction); comprehensive hypothesis testing; and independent repetition and confirmation. These are all necessary irrespective of content area, whether for "science" in the traditional sense or for any other facet of human awareness.

Science is not a field of knowledge; it *is* knowledge. The advancement of science is the enrichment of mankind. What we call "human progress" is quite literally the historical sum of innumerable scientific additions. Derived from the Latin *scientia* meaning knowledge, science, in its broadest sense, conceives most concepts and sculpts most objects. Science, today, is wondrous, and scientists, in a sense, are worshipped.

There is one area, however, where science is controversial, where inquiry is questioned and advancement criticized. Science in the service of

[1]See *Commercializing Defense-Related Technology*, edited by Robert Lawrence Kuhn (New York: Praeger, 1984).

national defense triggers hot debate. Some would say that scientists have the moral right to control the potential use of their personal creativity and the moral imperative to prevent their innovative output from producing weapons of war. This lofty position bespeaks high tone and laudable ideals yet is flawed fatally by inconsistency and illogic.

The simple reasoning framed for America, is thus: Such lofty positions can be espoused only in a free society; a free society will remain free only by military strength; military strength will be guaranteed only by state-of-the-art science. This is the real world. (Examples of free societies flourishing devoid of military strength? They only prove the point: All rely, at last resort, on the United States.)

National defense demands technological superiority. Parity in military science, for a nation without expansionist designs, is not good enough; equality just will not do — it's too close, a slight error and you're behind. And being behind is no place to be, not in this game, not with all the chips in the pot. In an electronic fairyland of blinking black boxes, where battlefield microprocessors command, control, and communicate, "leapfrogging" is the ever-present danger.

In past wars we could survive a slower tank or smaller sub, but in future encounters missing a scientific breakthrough in missile defense or sublocation technology could be disastrous. Our country is committed upfront: We will not be the aggressor. When the other side picks time and place, we had better field superior weapons and surer systems. When we concede quantity and number, we had better stress quality and expertise. The issue, of course, is more deterrence than triumph. We must *prevent* the next war, not win it.

Yet the world moves on. Subtle shifts redefine the nature of power. Today, well into the final fifth of the twentieth century, American security stakes out broader boundaries than ever before. More is encompassed within our vital needs as a nation. The economic thrust of Japan, for example, is a threat every bit as real as the military menace of the Soviets. Not the same, of course, but every bit as real. Computers and communications are also extending security boundaries. The profusion of information amplified by the ease of transmission lowers entry barriers for those with disruptive intent.

The battles of the future will be fought on vastly more complex terrain, contested more with ideas and products than with armies and navies. Confrontation among nations — attacks, provocations, insults — will assume new forms and novel shapes. Troop movement across Europe is virtually an anachronism — superpower nuclear standoff has seen to that. We must secure the standoff with military strength through technological supremacy, but that is not enough. An irrefutable defense capability, in the words of the logician, is "necessary but not sufficient" for national security.

This, then, is the *new* vision of national security, a broad concept embedding economic, social, education, cultural, and intellectual components as well as military ones, a concept increasingly being called "*Comprehensive* National Security."

Scientific superiority must maintain America's Comprehensive National Security just as it must assure the subset of preeminent military might. The first nation, for example, to mass produce future generations of integrated circuits — 256K, 512K, 1,024K — will capture high ground and strong position. The country that pioneers genetically enhanced food production will wield commanding influence in world politics, well in excess of Arab oil's peak power.

Comprehensive National Security must become our redeployed concept of self-protection. Mechanisms of competition, not machines of warfare, are now the critical concern. We must construct a *comprehensively* secure country, and American science is our primary building block.

Following is the domain of Comprehensive National Security, with each area evincing the central role of science.

Military: Maintaining technical superiority in weapons and delivery systems is the *sine qua non* of national security. Responsiveness, reliability and redundancy are also cardinal characteristics. American science should be proud to participate in sustaining freedom.

Economic: Strengthening the industrial base of the United States is a quintessential component of Comprehensive National Security. In past centuries countries could make up with military aggressiveness what they lacked in economic resourcefulness. This is no longer possible. Countries will survive and prosper or suffer and fall in direct relation to their production capacity and commercial acumen. The premier growth industries of the next decade — telecommunications, personal computing, biotechnology, and health care — are all science based. Scientists are not only involved in creating novel high-technology ventures but also in developing fresh approaches to traditional businesses. Both are prescribed for American economic health.

Social/Political: Structuring society for the benefit of all people is our contemporary megaproblem, labyrinthian in complexity, long term in solution. We must be able to meet our oft-stated goals of equality, opportunity, care, and concern for citizens of every age, sex, race, creed, religious belief, and so on. A populace well pleased is an intrinsic part of Comprehensive National Security. Though human systems are fiendishly more intricate than material systems, social scientists are as clever and inventive as their physical science counterparts. The use of sophisticated techniques in sociology, political science, and the like provide a core of hard data, certainly superior to the self-serving rhetoric of political palaver.

Educational: The minds of the young are the blueprints of the future. What we teach, and how they learn, will plot America's course — with the trajectory now being set in our schools. Science, here, contributes more than tools, though the personal computer will revolutionize both teaching and thinking. (Free enterprise has given the United States a jump of at least half a generation over the Soviet Union in acclimating children to personal computers.) Science teaches logic, how to use it, when to overrule it. It catalyzes enthusiasm for investigation and analysis; it teaches respect for proper rationale and confirmed proof; it offers the thrill of exploring unchartered areas, of using insight, of making discovery, of finding truth. Science replaces rote by rigor and memorization by reasoning. Science is no longer the exclusive domain of the elite; it is the language of all.

Cultural: The identity of a nation affects its cohensiveness; self-image determines self-confidence. Building American culture buttresses American security. Science, the complement of culture, supports its promulgation and propagation. Culture thrives on wide accessibility, and science provides the nutrients of transmission — television, radio, cable, satellite, video disks/cassettes, motion pictures, computer networks, and interactive video. Science has also fashioned marvelous techniques for enhancing effect, making culture more pleasurable and more veritable, conveying emotion, and making impact.

Intellectual: In the twenty-first century information will be the new medium of exchange. (Money, that archaic commodity, will be bytes in computer memories and numbers on computer screens.) International leadership will be framed in terms of cerebral skill not military prowess. A nation's prestige will be built by its intellectual endowment, not the number and size of its bombs and rockets. Scientists from all disciplines will contribute, from philosophy and astronomy to mathematics and music; new information will be prized, even from fields without direct economic benefit. Human values will have changed and human worth redefined.

A word, here, for *pure* science, basic research is the foundation of science, the platform for progress, the precursor of revolution. One cannot know in advance where seminal breakthroughs will come and what application technologies may be. Instinct and intuition, not program and project, are the requisite sources of energy. Basic research is a stimulant for creativity; it is, in all fields, an absolute necessity.

Sensitivity to scientists as well as appreciation of science is vital for optimizing national output. Scientists, by personality, are not easily coerced, not easily directed. Indeed, such is their strength. Scientists must be free to wander and explore, to confront blind alleys and to shatter tradition. Society must establish incentive systems to encourage scientists, giving them maximum motivation to imagine and construct. We must nurture and develop America's premier natural resource.

PART II

INITIATIVES FOR QUANTUM CHANGES

3

Initiatives for Transforming the American Economy

Eugene B. Konecci, J. R. Kirkland, George Kozmetsky, and Raymond W. Smilor

OVERVIEW

Initiatives for transforming the American economy are divided into three sections. These sections focus directly on coherent, cohesive approaches for utilizing technology venturing to help transform the American economy.

Section I deals with initiative areas for critical *national policies* required for quantum changes to position the United States to succeed effectively in a fiercely competitive global marketplace. These initiative areas are:

- Invest government resources productively to reshape America's industrial base.
- Assure America's scientific, technological and economic preeminence in a global economy.
- Form institutional coalitions between the public, private, and nonprofit sectors.

Section II treats initiative areas for actions and policies on the *state level* that foster economic and social development. These initiatives areas are:

- Identify and develop high-technology resources and capabilities.
- Forge collaborative relationships between universities and corporations for selected research and development (R&D) projects.
- Foster linkages between state and local government entities, the business community, and universities.

- Build local community activities and programs.

Section III focuses on initiative areas for *private-sector* leadership to take advantage of entrepreneurial opportunities in technology venturing. These initiative areas are:

- Develop imaginative ways to link public-sector policies and actions with private-sector investments.
- Commercialize technology venturing spin-offs.
- Develop creative and innovative management to strengthen America's competitive position in the new global economy.

SECTION I: CRITICAL NATIONAL POLICIES FOR QUANTUM CHANGES: DOMESTIC AND GLOBAL

The American economy is undergoing a basic transformation. It is moving from an older economy to a newer economy. The forces behind these changes have emerged from technology venturing, which encompasses newer institutional relationships, risk taking, and risk sharing.

The "old" American economy is becoming technogically and economically obsolete. There is a pressing need to develop a new economic infrastructure. Today's environment for change is fundamentally different from even a decade ago. The process of change has been dramatically accelerated. The "old economy" emphasized cheap and abundant natural resources, borrowing over savings, growth over efficiency, and quantity over quality.

The "newer economy" is reversing the old economy trends in a global context. Solutions to critical issues and problems now demand an integrated, holistic, flexible approach that blends technological, managerial, scientific, financial, socioeconomic, cultural, and political ramifications in a period of extreme time compression. Newer interinstitutional relationships are emerging in this time-compressed environment. These are resulting in changes to traditional approaches of risk taking and risk sharing.

The new economy is fostering new relationships between:

- The federal government and the private sector.
- The federal government and the state governments.
- State governments and the private sector.

- Universities and governments.
- Universities and the private sector.

These emerging relationships are seen most clearly in technology venturing. Technology venturing is the newer American response to the changing economy. It is a grass roots approach for strengthening the U.S. leadership role and its position in a global marketplace.

Technology venturing, when coupled with the emergent institutional changes, becomes a major driver for quantum change in dealing with many of the current problems besetting American society. A driver is a catalyst that alters the composition of society including its institutions and the way we choose and allocate resources. It has dramatic impacts on the nation, individual states, and local communities. It affects the viability of our industries, the growth and survivability of our business enterprises, and the role, scope, and purpose of our private/public-sector institutions, including higher education.

Initiatives for Critical National Policies

This section stresses the need for critical national policies fostering technology venturing among the federal government, higher education, and the private sector. Three current federally sponsored macro- or large-scale programs are discussed: (1) the strategic defense initiative program; (2) permanent space station and commercialization program; and (3) public infrastructure systems projects. These programs as well as other major federal projects can well become drivers that change and improve our domestic economy and provide a stronger base for U.S. foreign trade.

These programs have three common characteristics:

1. They require the collaborative efforts of government agencies and private firms to provide capital, services, and technology, as well as other resources in planning, development, and implementation.
2. They utilize state-of-the-art technology and scientific breakthroughs that have significant technology commercialization potentials.
3. They have far-reaching economic, sociocultural and environmental impacts beyond their immediate programmatic objectives.

Critical national policy initiatives and recommendations to help position the United States to succeed domestically as well as in a global marketplace are set forth in the following three parts:

Invest Government Resources Productively to Reshape America's Economy.

Assure America's Scientific, Economic and Technological Preeminence in a Global Society.

Form Institutional Coalitions among Public, Private, and Nonprofit Sectors.

Invest Government Resources Productively To Reshape America's Economy

Large-scale programs are major federal stimulants for the new economy. These programs can provide the leading edge investments for transforming our high-technology industry as well as our natural resources and basic industries. These federal programs in the next decade or two will require federal outlays of a minimum of $1 trillion to well over $4 trillion. From a critical national policies point of view, these expenditures must be evaluated in terms of not only deficits to the federal budgets but also longer-term capital investments. The commercialization of these capital investments requires that federal and state governments, businesses, and universities establish mutually beneficial coalitions on projects to make their own supplementary investments of capital, human, and intellectual resources.

In summary, Congress and the administration must formulate and establish policies that address the following initiatives:

Initiative 1
Develop cohesive and coherent policies to strengthen the U.S. competitive position globally.

Initiative 2
Extend major federal macroprograms to foster technology venturing.

Initiative 3
Consider newer forms of public risk sharing beyond the ability/capacity of the private sector or an individual state.

Initiative 4
Enact a National Commercialization Act.

Initiative 5
Broaden allocation of federal R&D funds both geographically and institutionally.

Initiative 1. Congress and the administration should review current, disparate, and inconsistent economic, and monetary policies that have an impact on American commercial and technological relations with foreign countries to develop cohesive and coherent policies to strengthen the U.S. competitive position worldwide.

Such policies should include but not be limited to:

1. Establishing monetary and technology policies that result in a reasonably valued dollar within stable and predictable foreign exchange markets.
2. Encouraging the development of international monetary programs for dealing with the debt of less developed countries. Measures are necessary to assure the integrity of the U.S. banking system and to provide capital to support trade with developing countries at competitive rates of interest.
3. Formulating a more stable and competitive interest rate structure.
4. Providing more flexible commercial and funding arrangements that surround large-scale technological undertakings. Such arrangements should encompass risks that are in the public interest.
5. Initiating the review of international technology venturing opportunities that are not counter-productive to the new economy and that enhance the U.S. competitive position worldwide.

Initiative 2. Congress and the administration should formulate and establish policies that provide for measures to extend major federal macroprograms to foster technology venturing for the creation of a new and modern American economic infrastructure. Such policies and measures should include but not be limited to

1. Extending the commercialization of "spin-off" products and services. These "spin-offs" must be identified in the early stages of the project development. Measures should be developed to make their commercialization timely. Measures should be developed to introduce "spin-offs" in a cohesive manner into domestic and international markets after reviewing appropriate risk taking and risk sharing measures for comprehensive security.
2. Modifying foreign licensing procedures. Procedures should include more than fees. They should contain arrangements that assure the creation and maintenance of American jobs upon commercialization of technology.

3. Considering appropriations to macroprograms as national capital investments. Accordingly, an accounting and management procedure should be established to treat such outlays in projects separately to increase public understanding that these are investments and not only expenditures in the traditional budgeting process.

4. Monitoring technology venturing developments and reporting on them in special ways such as through the State of the Union address. Topics for such reporting efforts should include newer employment opportunities, trade balances, benefits from licensing, status of emerging industries, newer products and services, revitalization of basic industries, increased productivity, needs and quality of the public infrastructure, and the competitive state of the defense and space industries.

Initiative 3. Congress and the administration should consider newer forms of public risk sharing for those critical and crucial needs that are beyond the ability/capacity of the private sector or individual state. Such policies and measures should include but not be limited to:

1. Reviewing the infrastructure needs of America with a view to develop a comprehensive public works program. Such a program is essential to strengthen American commerce and enhance America's international trade position.

2. Analyzing the "scaling up" requirements of large-scale projects to develop appropriate policies for fostering the linkages necessary to accomplish such projects when the inherent risks are abnormally large for the private sector.

3. Examining resource needs of the United States, by region, to determine long-range requirements for water, air quality, and general quality of life.

4. Identifying replacement needs of global infrastructure systems and developing strategic approaches for U.S. redevelopment of these systems.

5. Reviewing policies for critical materials to assure resource and military viability especially in regard to development of advanced scientific research and newer technologies.

6. Providing risk insurance for large-scale projects conducted in developing and Third-World countries.

7. Ensuring abundant and reasonably priced energy supplies, especially in regard to the development of new electrical generating capacity.

Initiative 4. Enact a National commercialization Act.

Measures for a National Commercialization Act should include but not be limited to:

1. Assisting state governments to foster technology venturing for both risk taking and risk sharing.

2. Providing measures that assure endowed financial resources with adequate annual income to assist in the growth of emerging industries and to undertake high-risk ventures.

3. Promoting and expanding active technology transfer between the Federal Laboratory Consortium and the private sector.

4. Fostering cooperation between public and private research performers and users, especially universities and industry, to encourage the commercialization process.

5. Studying the need for legislation to establish an Advanced Technology Foundation (ATF) as an independent civilian funding agency that fosters the innovative use of new federally supported technology for a wide spectrum of basic and emerging industries as part of the commercialization process.

Initiative 5. Federal agencies should adopt a policy to broaden their allocation of R&D funds both geographically and institutionally.

Policies and measures should include but not be limited to:

1. Encouraging the more rapid development and expansion of research/teaching in private and public higher educational institutions in all 50 states through appropriate grants and funding.

2. Broadening the research and development support. Expanding our scientific and technology base should include developing newer "lightning rod" institutes and centers to supplement centers of excellence and to meet the new economy requirements for scientific, engineering, and managerial human resources.

Assure America's Scientific, Economic, and Technological Preeminence in a Global Society

Key elements assuring U.S. preeminence in a global context are the following:

- U.S. scientific, technological, and economic preeminence depends on the strength, vigor, and dynamism of our unique entrepreneurial American culture and heritage.
- The United States does not need to adopt the Japanese style of "targeting" processes for economic preeminence.
- Scientific and technology education from primary through higher education needs to be strengthened within the current state and local infrastructures.
- A more cohesive approach is necessary for the encouragement of U.S. high-technology venture business.

Technology venturing is a catalyst for encouraging the emergence of a myriad of technology venture businesses in the context of a private enterprise system that has always been the unique American way to achieve and maintain U.S. scientific, technological, and economic preeminence. America's strength has always been our ability to be scientifically creative, technologically adept, managerially innovative, and entrepreneurially daring.

Congress and the administration must formulate and establish policies that address the following initiative areas:

> **Initiative 6**
> Establish policies that achieve and maintain U.S. preeminence in selected areas of international scientific competition.
>
> **Initiative 7**
> Develop policies for utilizing technology as a national resource to accelerate its commercialization for U.S. economic preeminence.
>
> **Initiative 8**
> Encourage and establish more cohesive measures to foster U.S. technological preeminence.

Initiative 6. Congress and the administration should encourage and establish policies that achieve and maintain U.S. preeminence in selected areas of international scientific competition and that can help to foster the new economy. These policies should include but not be limited to:

1. Promoting basic scientific research by discipline as well as by associative interdisciplinary scientific research that encompasses engineering

research and all other educational linkages that are necessary to meet the longer-range goals for the new economy.

2. Making key scientific research capabilities available (by remote access, if appropriate) to a broad spectrum of academic research. More specifically, the supercomputer is essential for extending knowledge in a broad spectrum of scientific, engineering and managerial research. The cost of the supercomputer should be matched with state and private funds.

3. Supporting research institutes and "lightning rod" centers at universities for cross-disciplinary work in basic science as well as engineering to support new economic infrastructures.

4. Achieving public understanding of basic constituents of matter and energy including high-energy physics to maintain an active and influential U.S. scientific position internationally.

5. Continuing vigorous and comprehensive R&D programs in space sciences, applications and technology development for U.S. preeminence.

6. Establishing a foresight and early warning reporting system by nation and discipline that assists in monitoring the state of scientific preeminence in the U.S. as well as our competitive position internationally.

Initiative 7. Congress and the administration should study the feasibility of developing policies to utilize technology wherever it may originate as a national resource and to accelerate its subsequent diffusion and commercialization for U.S. economic preeminence.

These policies should include but not be limited to:

1. Establishing mechanisms for delivering added benefits to the American people through commercializing basic and applied research as well as program and project developments by all performers of federally sponsored R&D.

2. Setting up a technological knowledge base: that is, a system for gathering information and knowledge on technological advances both domestically and internationally.

3. Establishing an international cooperative organization or consortium of highly industrialized nations and Third-World nations to hold conferences that discuss exchanges of high-technology information, impacts of future national policies for microelectronics, robotics, and other high technologies, and ways for increased direct investments, joint ventures and joint research projects. Conferences could explore the application of U.S. technology for European economic restoration and for modernization and industrialization of Third-World nations.

4. Evaluating mechanisms for *de novo* establishment of international standards for emerging high technology that encompass function, quality and environmental specifications within a global, competitive system.

Initiative 8. Congress and the administration should encourage and establish more cohesive measures to foster U.S. technological preeminence.

These measures should include but not be limited to:

1. Authorizing the establishment of a permanent R&D tax credit of at least 25 percent.
2. Extending R&D tax credits to include depreciation of research and advanced development equipment and facilities.
3. Permitting corporate tax deductions for state-of-the-art science and technology equipment and contributions to all educational levels.
4. Strengthening intellectual property protection including patents, trademarks, copyrights, and inventions for those in both public and private institutions.
5. Broadening R&D partnership tax deductions to include costs incurred in market research and development, especially for emerging products and services.
6. Initiating a study across a broad spectrum of technologies to determine appropriate depreciation and other tax policies that will increase our ability to maintain U.S. global technological and economic preeminence. Special consideration should be given to depreciating manufacturing equipment and facilities in a single year because of technology risk and obsolescence.
7. Establishing financial and manufacturing free trade zones.
8. Assisting large and small businesses in exporting U.S. products and services through up-to-date trade and market data.

Form Institutional Coalitions among the Public, Private, and Nonprofit Sectors

New institutional arrangements should be created among various sectors of the new American economy.

- Regulatory problems combined with a lack of national policies hinder required coalitions for utilization of scarce U.S. scientific and technological resources necessary to meet global competition in an appropriate time frame.

- There is a need to provide measures that foster private sector coalitions for public infrastructure programs that cut across states and regions.
- Currently, long-term federal macro- or large-scale programs focus on provisions for forming research and development coalitions, especially with friendly foreign, competitive nations. There is minimum emphasis on subsequent commercialization arrangements.
- There is a need to develop better means and methodology for the management of large-scale programs.
- Better communication between large-scale project managers and scientists is necessary to develop newer principles and technologies for improving productivity in public infrastructure programs.

The need for intersectional harmony and collaboration arises from the nature and type of international competition that confronts the United States.

America's future depends on our abilities to form essential institutional coalitions that can support and diffuse the newer technologies. These technologies can become our nation's growth industries that will help to stimulate a robust new economy. More cooperation is necessary among those institutions that perform basic research, those that do applied and developmental research, those that are the source of R&D funding, and those that perform subsequent commercialization. These institutional arrangements, by permitting more flexibility and adaptability in assessing and implementing research and technology, can assist in meeting the nation's needs in resolving rapidly changing security, socioeconomic, and political issues.

Congress and the administration must formulate and establish policies that address the following initiative areas:

Initiative 9
Reassess antitrust laws and guidelines to bring them up to date with the realities of a new U.S. economy.

Initiative 10
Evaluate the effect of international joint ventures and coalitions in high-technology industries and large-scale projects.

Initiative 9. Congress and the administration should consider polices and measures that reassess antitrust laws and guidelines to bring them up to date with the realities of a new U.S. economy.

These policies and measures should include but not be limited to

1. Evaluating the reduction of barriers to joint R&D ventures.
2. Considering the impact of current antitrust legislation on U.S. protectionist policies.
3. Reexamining the concentration guidelines for basic and high-technology industries in light of current and future global competition.
4. Developing measures to encourage research that evaluates the efficiency of American industries as well as the size of supplier and user firms relative to antitrust requirements and global competition.
5. Assessing the unintended consequences of current antitrust legislation.

Initiative 10. Congress should fund and the administration should appoint a commission to evaluate the effect of international joint ventures and coalitions in high-technology industries and large-scale projects on the competitive position of U.S. firms in the international marketplace.

This evaluation should include but not be limited to:

1. Monitoring joint ventures and their impact on American industry competitiveness, job creation, and economic preeminence.
2. Assessing the role, scope, and effect of foreign governments in coalitions between countries and among private, public, and nationalized firms.
3. Assessing the impacts of joint ventures on infant technology-based industries.

SECTION II: STATE ACTIONS AND POLICIES FOR ECONOMIC AND SOCIAL DEVELOPMENT

State governments are taking a leadership role in technology venturing. They have initiated a variety of actions and policies to foster economic and social development through technological innovations. They have become a major factor in assisting the transfer of technology to the market.

Bellwether states such as California, Massachusetts, Florida, North Carolina, Texas, and New Mexico are:

- Moving to aid the growth of high-technology industry.
- Advancing the development of indigenous companies for job creation and economic diversification.
- Reassessing and reshaping the educational structure to meet the needs of a changing society.
- Encouraging advanced R&D activities and promoting innovation.
- Assisting local communities in economic and social development.
- Participating in public/private coalitions and private partnerships.

State governments have just begun to recognize that technology is a resource to be used. Like a catalyst, it can be a major stimulant for new business formation, industrial innovation, and employment growth. It can be adapted to and utilized for each state's individual needs. By using technology as a resource, states are building a dynamic and cohesive approach to technology venturing. They are shaping tomorrow's private/public infrastructure. Through collaborative efforts, they are encouraging entrepreneurial activity and helping to balance the different requirements of business, education, and the public.

Initiatives for State Level Actions and Policies

A series of developments has led state governments to transform their traditional approaches to economic development. They have begun to initiate dozens of programs, upgrade schools of science, mathematics, and engineering, establish research parks, appoint special advisory bodies to promote industrial innovation, and reassess their educational programs and incentives for fostering technology growth.

The following factors for the transformation at the state level evolved:

- The realignment of and the changing responsibilities among federal, state, and local governments.
- Declining federal government funds for basic research.
- New pressures for job creation and employee retraining.
- Rapidly increasing demands for scientists, engineers, and technicians.
- Declining usefulness of state zero-sum industrial "piracy" strategies.
- The need to encourage the export of state products and the attraction of foreign companies.

State responses to these factors have been admirably collated by the National Governors Association (NGA) in its "Final Report of the NGA Task Force on Technological Innovation." The NGA report comments on state high-technology activity in five broad program areas:

1. Policy development.
2. Education, training, and employment.
3. Linking university/industry research.
4. Technical and management support.
5. Economic support and incentives.

The next stages of economic and social development at the state level must address the creation of cohesive and coherent state level actions and policies in the following four initiative areas:

Initiative 11
Identify and develop high-technology resources and capabilities.

Initiative 12
Forge collaborative relationships between universities and corporations for selected R&D projects.

Initiative 13
Foster pioneering linkages among state and local government entities, the business community, and universities.

Initiative 14
Build local community activities and programs.

Initiative 11. The state governor and legislature should develop actions and policies that identify and develop high-technology resources and capabilities.

These actions and policies should include but not be limited to

1. Establishing within a collaborative framework an advisory board or council with members representing business, government, academia, labor, and other important constituencies for analyzing and reporting on state capabilities and needs to promote technology research and development and industrial expansion.

2. Appointing a task force or commission to:
   a. Determine critical issues and needs that will influence the state's longer-term future economic growth and development.
   b. Receive input from public, private, academic, government, and other sectors of the state.
   c. Study each of the needs and issue areas in depth and develop appropriate recommendations.

3. Setting up a task force or commission to identify needs for all levels of education for the state's long-range infrastructure to:
   a. Assess educational requirements and make recommendations.
   b. Promote statewide science and technology education as well as basic and applied research and development.
   c. Evaluate current and future state needs for teacher education and training.
   d. Determine the need for high-technology centers.
   e. Institute customized worker-training and retraining programs to assure and adequate supply of skilled and highly trained employees to meet the demands of defense and nondefense technology-based industries and services.
   f. Evaluate the adequacy of scientific, engineering, and technology management higher education for entry and postentry needs.
   g. Determine the requirements for establishing links among academia, industry, and financial entities for high-technology "start-up" and "take-off" companies.

4. Evaluating the role and scope of current state agencies responsible for economic planning and development to:
   a. Study the feasibility of expanding their current responsibilities or refocusing their efforts to promote technology and industrial innovation.
   b. Establish separate entities for technology development including state technology partnership funds for centers operated by consortia of business, labor, financial institutions, academia, and local development districts with funding from both public and private sources.
   c. Create a state technology park corporation to establish a statewide network of facilities dedicated to specific areas of science and technology with funding from public, private, and academic sources.

d. Establish nonprofit, quasi-public institutions to oversee high-technology development including science parks, research parks, R&D crescents, corridors, highways, and triangles.

5. Enacting a State Commercialization Act to create a revocable trust to award seed money to eligible joint venture enterprises involved in longer-term, high-risk, technology-based developments.

6. Evaluating the adequacy of the state's capital venture industry as well as the availability of entrepreneurial talent to develop required growth of technology-based industry.

Initiative 12. The state governor, legislature, and boards of regents should forge collaborative relationships between universities and corporations for selected projects for the long-range technological and economic development of the state.

These actions and policies should include but not be limited to:

1. Encouraging broader university and industry cooperative participation through corporate gift and contract research giving.

2. Studying means of broadening collaborative relationships between universities and corporations in targeted technology areas that are significant to the future of the state. In this regard, alternatives should be evaluated for establishing and utilizing institutes of excellence as well as "lightning rod" centers for special technology areas. Financial support can come from the private sector with public sector incentives including matching, set aside of a small percentage of current special fees, or taxes, land, etc.

3. Studying the need to modify state antitrust laws in order to encourage cooperative corporate or trade association research.

4. Determining the need, location, timing, and methods of linkages for establishing research parks. Such analysis should clearly delineate the needed capital investments, appropriate public/private sector ties, and required maintenance and replacement costs.

5. Considering the feasibility of allowing universities to retain all or part of their overhead generated from corporate and federally supported research as an incentive for fostering university R&D and as a reinvestment for advanced technological developments.

6. Facilitating the administration of collaborative university and corporate research including modifying appropriate state/academic policies to encourage sharing among all parties in the subsequent benefits of the research.

Initiative 13. The state governor and legislators should foster pioneering linkages among state and local government entities, the business community, and universities.

These actions and policies should include but not be limited to:

1. Encouraging the establishment of a variety of areas for technology innovation and industrialization including research and development parks, innovation centers, incubators, and technology commercialization centers.
2. Establishing technical assistance centers for smaller technology-based businesses, including appropriate academic institutions serving as links to technical, management, financial, and educational resource capabilities in the state.
3. Promoting the establishment of training and support programs for entrepreneurial developments critical for technology commercialization.
4. Developing special incentives for technology diversification such as tax incentivies, enterprise zones, direct financial assistance in the form of low-interest bank loans and guarantees as well as start-up capital needs of new technology firms.
5. Encouraging the formation of joint research ventures by pooling research talent from companies to maintain the state's preeminence in technology.
6. Fostering the formation of academic-corporate consortiums to develop required methodologies for large-scale technological programs to support the state's participation in federal macroprograms.
7. Promoting the formation of interstate academic consortiums to utilize limited research facilities and equipment, e.g., supercomputers.
8. Encouraging the utilization of in-state federal government laboratories through the establishment of state incubation and innovation centers linked to the laboratories.

Initiative 14. Mayors and other locally elected officials should foster collaborative efforts to build local community activities for specific technology venturing programs.

The actions and policies should include but not be limited to:

1. Developing the goals or longer-range plans for the community by establishing an appropriate task force represented by all critical elements in the community. When necessary, the scope should be expanded to

include other communities that comprise natural crescents, corridors, triangles, or technopolies.

2. Encouraging the creation of regional academic, corporate, and community organizations as well as entrepreneurial activities that meet the goals and objectives of the community.

3. Providing the requisite leadership role for assuring the collaboration of other public/private groups and organizations to achieve the community's common goals in a cohesive manner.

4. Assuring that the community utilizes the full extent of the federal, state, private, and local resources available for technology venturing.

5. Encouraging the development of indigenous companies within the community.

6. Initiating the development of long-term, high-risk, technology-based efforts as well as large-scale programs for the civil infrastructure to utilize the resources provided by the State Commercialization Act.

7. Fostering the development of local business development firms or capital venturing to attract and retain start-up opportunities available from local incubators and innovation centers.

SECTION III: PRIVATE-SECTOR ENTREPRENEURIAL OPPORTUNITIES FROM TECHNOLOGY VENTURING[1]

The pursuit of research and development does not necessarily require that those who provide sources of funds or those who perform the R&D bring the innovations to the marketplace. Research can be performed on demand; innovation cannot. Innovations, therefore, are entrepreneurial opportunities.

American business leaders and entrepreneurs are the catalysts in the utilization of innovations as resources to create the new American economy. Their leadership choices determine which technology is to be commercialized, when it should be taken to the market, and which geographic locations will do the research, development, manufacturing, and marketing. These choices will create a new American economic base.

The collaborative process under technology venturing can be accelerated when private-sector leadership works in conjunction with federal, state, and local governments. The commercialization of technology is an

[1]See *Corporate Creativity: Robust Companies and the Entrepreneurial Spirit,* edited by Raymond W. Smilor and Robert Lawrence Kuhn (New York: Praeger, 1984).

entrepreneurial process that can shape our new economy. But we need to understand this process better. Through technology venturing, small, medium, large, and giant business firms can utilize emerging technologies, forge newer institutional relationships, become more competitive, develop indigenous companies for job creation and economic development, encourage advanced R&D activities, help reshape the educational structure in training and upgrading workers, and assist local communities in economic and social development.

American business leadership opportunities are forthcoming in the new economy from a series of stimulants. These stimulants include:

- The federal government is initiating a quantum change through a series of macro- or large-scale programs that extend over a decade or more. These programs utilize both high and low technology that have subsequent innovative domestic and worldwide commercialization potentials.

- Many states have taken actions and initiatives to build newer infrastructures and encourage entrepreneurial efforts.

- Research/teaching academic institutions have initiated imaginative sustainable longer-term collaborative relationships with the private sector.

- Newer linkages have been formed between government, universities, and the private sector, which are resulting in research parks, developmental corridors, crescents, and triangles, and high-technology highways.

- Emerging joint research efforts are bringing together research talents from rival companies as an answer to foreign competition.

- Capital venture participation is affected by changes in fiscal policies and tax laws.

- Creative and decentralized institutional arrangements among the public, private, and nonprofit sectors are providing for basic research, promoting economic and cultural advances and improving quality of life in more communities.

- Other nations' avowed goals for scientific and technological competition have helped to refocus American needs for preeminence in science, technology, and commercialization.

- There is a growing realization that extremely large world markets exist in large-scale technologies for the public infrastructure in developing and underdeveloped nations.

Action Initiative for Focusing Private-Sector Leadership on Technology Venturing Opportunities

In the past, for a number of good reasons, private-sector leadership has been focused on managing adversarial and advocacy issues. Technology venturing has shifted the focus of business to more collaborative, coherent, and cohesive approaches and actions.

Although technology venturing holds much promise for the future, it is still in its formative stages. All parties that comprise such efforts have yet to develop fully and understand completely all the relevant interrelationships. The next stages for the private sector are to take necessary actions to build positive relationships with the government and academic sectors to identify and extend technology venturing in the following three initiative areas:

> **Initiative 15**
> Develop imaginative ways to link public-sector policies and actions with private-sector investments.
>
> **Initiative 16**
> Commercial technology venturing spin-offs.
>
> **Initiative 17**
> Develop creative and innovative managment to strengthen America's competitive position in the new global economy.

Initiative 15. Private-sector institutions should develop imaginative ways to link public-sector policies and actions with private-sector investments.

Actions and activities should include but not be limited to

1. Creating an Academy of American Industry (AAI). This academy would conduct studies and research for the effective development and

growth of American industry in the changing world economy. AAI would collect information, monitor and analyze changes, identify critical new applications of sciences and technologies, and make recommendations that increase, maintain, or create domestic industries for global markets.

2. Supporting the American Society for Macroengineering. The society provides a forum for the identification, study, and discussion of the many issues related to the development of macroengineering projects, programs, and systems. These issues involve and interact with many other concerns of industry, government, and the public at large. The society's mission is to increase public awareness and understanding of these issues, promote public sensitivity to macroengineering projects and programs, and address related public policy issues. The objectives of the society are to:

a. Help create a climate conducive to proper planning, approval, and execution of exceptionally large and complex engineering/construction projects.
b. Promote the conception and implementation of such projects.
c. Help generate and disseminate the wide range of basic information required for such projects.
d. Provide a forum in a noncommercial atmosphere for open discussion and analysis of the problems inherent in such projects.
e. Foster interprofessional, interdisciplinary, and intersectional interest in macroengineering issues.
f. Educate the general public as to the challenges, advantages, and impacts of such projects.
g. Stimulate the performance of required research.
h. Enlarge public consideration and review of such projects so that consensus support is maximized.

3. Establishing a number of nonprofit Institutes for Large-Scale Programs (ILSP). These institutes can be consortiums of university and public- and private-sector institutions. ILSP would direct its attention to problems and issues dealing with very large-scale programs that utilize technology as a resource.

Initiative 16. Private-sector institutions should commercialize technology venturing spin-offs.

Actions and activities should include but not be limited to:

1. Establishing a clearinghouse of technology opportunities provided by national policies and programs, state actions and policies, and community efforts and initiatives.

2. Fostering joint meetings with business associations, professional societies, and trade organizations for the dissemination and utilization of incubator and innovation centers.

3. Utilizing academic research and talent.

4. Supporting financial initiatives for start-ups, capital venturing, and business development.

5. Taking advantage of federal laboratories technology development and other research and development supported by NASA, DOD, DOE, NSF, and other government agencies.

6. Initiating technology venturing on an international scale. These ventures can include foreign government research institutions, academic centers, and corporations.

7. Assisting in the commercialization of technology developed in national federal laboratories for both small-scale and large-scale innovations for rebuilding basic industries and developing new industries.

Initiative 17. Private-sector institutions should develop creative and innovative management to strengthen America's competitive position in the new global economy.

Actions and activities can include but should not be limited to:

1. Identifying needs and establishing support for future research. Example research needs identified are the steel industry and public infrastructure systems including water, construction, pollution, and transportation.

2. Identifying and eliminating barriers to technology venturing at federal, state, and local levels. These may include tax incentives for modern research and development facilities, antitrust for joint industry research, and other national and state regulations.

3. Establishing ways to provide meaningful reporting of accurate information to appropriate authorities on the impact of technology venturing investments on the private sector.

4. Recommending incentives to accelerate technology venturing with the appropriate background materials, e.g., 25 percent permanent R&D tax credit capital gains law incentives on emerging industries growth, tax changes on capital distributions, and employment by company size and geographic location.

PART III

NATIONAL POLICIES: STRATEGIC IMPLICATIONS

4

Technology Venturing: A Visionary Challenge for Prosperity, Security, and Opportunities

The Honorable Lloyd Bentsen

Back in a simpler time when a confident, energetic America was flexing its muscles to take up the challenge of world leadership, a candidate for vice-president told members of the Hamilton Club in Chicago, "It is far better to dare mighty things, to win glorious triumphs — even though checkered with failure — than to take rank with those poor spirits who neither enjoy much nor suffer much because they live in the gray twilight that knows not victory or defeat."

Eighty-five years later, the wisdom of Teddy Roosevelt has special meaning and significance. As risk takers who set the tempo for our free enterprise systems, technology ventures are the inheritors of the uniquely American spirit symbolized by our twenty-sixth president. Some have won glorious triumphs, and others have known failure — but none has chosen to hide in the gray twilight and mediocrity.

It's a very special pleasure for me to contribute to this volume. I spent 16 years of my life in the private sector, building up a business, taking the risks, raising the capital, and creating the jobs and opportunity. I know wheat it means to be an entrepreneur — and what that spirit means for Texas and for America.

As a member of the Finance Committee, the Joint Economic Committee, and the Joint Tax Committee in the Senate, I spend a lot of my time in Washington dealing with the tough economic policy questions that are so important to our ability to compete and prosper. In those deliberations I try to be an advocate for America's entrepreneurs — our people with ideas, courage, and the will to succeed.

What can government do to broaden the opportunities for growth and entrepreneurial success? What can all of us — the public and private sectors, business and labor, liberals and conservatives — do *together* to restore a drive, a lift, and a vision to the American economy.

Lately, there has been much talk about industrial policy, which is the latest in a long line of snake oil remedies claiming to cure everything that ails the American economy.

If we have learned one thing during a very difficult decade of inflation, stagnation, and recession, it is this: There are no easy answers. We've tried them all, and they don't work. We've primed the pump and pumped the prime. We've tried Laffer curves and supply side economics.

I think it's time for America to understand that if we want to prosper and compete in the decades to come there is simply no alternative to hard work in the context of sound economic and fiscal policies.

Here is my suggestion for an American industrial policy. Forget the planning and the formulas. Forget the allocations and memories of the National Recovery Act. If there is to be an industrial policy, let it be simple and let it concentrate on growth. Let it say that the job of government is to create an environment in which the private sector can grow most effectively. When the obstacles to growth have been eliminated, government should get out of the way.

It just doesn't make sense for America to try to clone Japan's economic success. Some people think that if we have our own MITI — our own National Development Board — our problems will be solved. They tend to forget that membership on those boards is often made up of people who are wedded to the *status quo*, which is the enemy of innovation and creativity. People who look to Japan's MITI as some secret of success also forget that Japan has had its fair share of misses, such as ship building and the steel industry. Today there is evidence to suggest that MITI's problems are starting to compound.

Here in America we don't need a tinkering, a meddling, regulating army of bureaucrats telling us where to put our resources. All we ask is a climate favorable to growth that gives us the freedom to succeed.

Technology venturing is the leading edge of the frontier of tomorrow. Successful business men and women understand the difference between easy, inflationary policies of simple expansion and tough choices that produce real growth. You know that real growth demands increased productivity; it thrives on stability, savings, and investment. Real growth is ignited by the spark and daring that is the genius of the entrepreneur.

A few years ago, if one were to name the main obstacle to real growth in the American economy, the reply might have been "inflation, fueled by high government spending." And that would have been right.

Today, if I were to pose the same question, the answer would be different. The roadblock that looms up before us today is a $200 billion budget deficit that threatens to destroy all our hard-earned economic progress.

Why is a $200 billion deficit an obstacle to growth? Because it places huge strains on our credit markets; it creates upward pressure on interest rates; it overvalues the dollar and makes it more difficult to export.

The danger of the deficit is becoming more obvious every day. It is poisoning our economy with uncertainty. The problem cannot be shrugged off or ignored. And it should not become a partisan political issue.

Under current spending and tax policies, $200 billion deficits are almost inevitable for the next five years even assuming the best economic circumstances. The cost of servicing the debt will grow by almost 16 percent a year. By 1989, almost half of all our personal income tax payments will go just to pay interest on the national debt.

That kind of dangerous deficit environment is the archenemy of growth. Today the first obligation of any government committed to growth must be to deal honestly and effectively with the problem of the deficit.

It is not enough to call for cuts in nondefense discretionary spending. Those cuts have already been made. You could eliminate nondefense discretionary spending entirely — you could zero it out — and you would still not balance the budget.

The easy answers will not work. The truth is that there are only three areas where spending is increasing: entitlements, defense, and interest. You can't cut interest payments by fiat, so that means we must take a close, hard look at defense and entitlements. And there's no sense fooling ourselves; we can't just cut spending and eliminate waste. We will need some new revenues, which is a polite way of saying more taxes, if we want to reduce the deficit. Hopefully, any new taxes will be the kind that encourage the savings and investment our economy needs so desperately.

Now, when we start talking about higher taxes, when you consider changes in the entitlements program and consider slowing down increases in defense spending, all of a sudden you're into some of the most controversial, sensitive areas of politics. That's why I think we need to make these tough choices together. Not as Republicans and Democrats, not as Congress and the administration, but as Americans working to build a future for our country.

With enlightened, courageous leadership, government can be the catalyst that brings us together to solve problems and pave the way for growth. Instead of pestering the private sector with regulation, government can instill a sense of fairness and willingness to sacrifice. Government can provide incentives for the economic behavior we want to encourage.

Let me give an example. I think our government should be encouraging high-technology, high-growth companies. We need high technology to compete; we need high technology to create new markets and new jobs; we need high technology to survive. The day this country loses its position of world leadership in technology and innovation, we place our freedoms and our future in jeopardy.

Some people simply *assume* that America will always be preeminent in science and technological innovation. These people are dangerously mistaken. They are whistling in the dark. They forget that research and development provide the bedrock for high-technology success. And today America is neglecting this groundwork. As a share of the gross national product (GNP), we are devoting 20 percent less to research and development today than we did in the 1960s. West Germany spends 2.15 percent of its GNP on civilian research and development (R&D); the figure for this country is only 1.66 percent, and that willingness to pay the price for R&D is one important reason why West Germany's productivity increased so much faster than ours during the 1960s and 1970s.

I don't have to tell technology venturers that America is locked in a continuous, high-stakes race with the rest of the world to bring new products to market. Any U.S. company that fails to advance technologically will soon discover its products have been rendered obsolete by competition from abroad.

Now it is true that much of our overseas competition enjoys unfair, artificial advantages like protection or subsidies. We cannot ask our people to compete at a disadvantage. We must insist that our trade agreements be strictly enforced and that our goods enjoy equal access to foreign markets. Our trading partners, the Japanese in particular, must understand that our patience has it limits.

But America's high-technology risk takers are not looking to government for favors or handouts. All they demand is equal terms of competition and a tax code that encourages economic success. America's prosperity in the years to come will depend in large part on a growing, dynamic, competitive high-technology sector. And that kind of high-technology performance demands a vigorous R&D effort. This is why I have helped write legislation and lead the effort to make the R&D tax credit permanent, to make depreciation of research equipment eligible for R&D credits, and to extend those incentives to start-up corporations.

I am especially alarmed at some of the trends we are seeing in basic R&D. Back in the 1960s, we were spending 7 cents of every research dollar on basic R&D. Today that figure is down to 3.6 cents. University research has been a major victim of this myopia. About 60 percent of our basic research is done at universities, but today industry contributes only 3 cents of every R&D dollar that comes into the universities.

I know that basic R&D is risky. I know it is expensive and takes years to generate a payoff — at a time when companies are under constant pressure to show short-term, bottom-line results. It's easy and convenient to mortgage the future and skimp on basic R&D. But, in the long run, it's a false economy that gives the whip hand to our competition.

As part of the effort to encourage corporate funding of basic research in America's universities, I have proposed offering a 25 percent tax credit for such expenditures that exceed historical levels. That's the kind of incentive that makes sense for the company doing the investing, for the university doing the research, and for the country determined to remain competitive.

Our universities have always been a key element in America's high-technology success. One of the main reasons Texas is fast becoming a mecca for technology is the existence of excellent research facilities at our universities. But looking at the country as a whole, it is alarming to learn that there are some 2,000 vacancies in our engineering faculties. The Japanese produce 163 engineering graduates per million population; the Soviet Union produces 260; here in the United States we produce only 63, and that statistic has frightening implications for our future ability to compete.

The shortage of qualified teachers is only half the story. Most of our engineering and vocational education schools are also working with outdated equipment. They're trying to teach the technology of the 1980s with the equipment of the 1960s.

Incentives for increased spending on basic R&D will encourage the kind of funding that should enable universities to attract and keep qualified faculty. As part of the effort to cope with the problem of outdated equipment, I have also proposed generous tax deductions for corporate contributions of state-of-the-art scientific and technical equipment.

John Danforth and I have worked hard on this legislation. We call it the High-Technology Research and Scientific Education Act of 1983. We know it's not going to solve all our problems, but we do think it is a step in the right direction that merits support. We think it is the kind of approach that works best for this country. We've identified a need — more

R&D, more university research, more engineering faculty, more modern equipment to lean on — and then we've provided incentives for the private sector to help solve the problem.

All too often we see government and the private sector working at loggerheads, fighting one another, distrusting one another. America can no longer afford the luxury of distrust and confrontation. All around the world our competitors are working hand-in-hand with their governments to innovate, capture markets, and create jobs. Here in this country, if we can replace confrontation with cooperation, I am confident we can build an America that can look with confidence to the challenges of the future.

Not long ago, there was a football coach in Washington who liked to say that "the future is now." The *future* ability of American businesses to compete and export and innovate will depend on what we do *now* — today — as a nation. Can we muster the courage to take on the tough issues and clear a path for growth in our economy? Can our public and private sectors join together and provide incentives for success? Can we curb the bureaucrats and encourage the entrepreneurs?

I think we can. I'm optimistic about America. After a decade in the doldrums our economy is recovering. If we can seize this moment to work together and fashion policies that sustain real growth, America can prosper as never before. We can achieve our destiny of greatness.

Many years ago, Daniel Webster challenged America to "develop the resources of our land — call forth its powers, build up its institutions, promote all its great interests — and see whether we, in our day and generation, may not perform something worthy to be remembered." We must respond to that challenge by producing products or services or ideas worthy to be remembered. Risk takers and leaders of the private sector have offered a vision of success for America. Now we are being asked — we are being challenged — to help find the will and the way to turn that vision into a reality of prosperity, security, and opportunity for all our people.

5

Technology Venturing: Collaborative Efforts

Admiral Bobby Ray Inman

I write about the future, about what I think will be a very bright one. But there are many challenges: There are potentials for losing our way, there are genuine concerns on the national security side, there are constant problems. But they are all manageable.

One looks at this country's great role of leadership. So many factors have gone into it — the system of government, the basic attiudes of the population — but it's been America's economic strength that has made the difference. That economic strength has depended in large measure on the speed in which we can move technology into the marketplace.

There have been periods of time when the United States has been brilliant at it and periods of time when we have not been very good. If one only focused on books and articles published on the Japanese success over the last ten years, you would think a new master race had come along and discovered whole new approaches to marketing. The fact is that a very disciplined nation has proven that *American ideas* can be put into effect for enormous economic benefit. Because throughout the 1960s and 1970s, Japan progressed largely on ideas they took from this country, ideas they used with great effectiveness: quality control, new approaches to production, and technology development.

If these were our only concern, the ability to catch up quickly would be easily described. But the world out there is changing and now new technology is being created abroad. It is being created by more innovative approaches to using talent. Japan, for example, is much faster at taking new technology and moving it to the marketplace; they are intense competitors as they move out to deal with the rest of the world.

Collaboration among the federal government, the university world, and the private sector had enormous impact in World War II taking us to unquestioned preeminence in the creation of new technology and in its commercial use. If one looks at what went wrong in the 1960s, there are a lot of reasons, but clearly losing that collaboration, including even between academia and industry, contributed to our slowing down in creating new technology and allowing the rest of the world to whittle away our one-time absolute supremacy.

It isn't chauvinism that drives my interest. It is the longer range view of creating jobs, of improving our standard of living, and of assuring our national security. I remain persuaded that our ability to prosper for these last 35 years has been derived not only from our military strength but from our economic vitality, particularly the degree to which we were at the leading edge of developing technology. It is an effort to define that leading edge that brings me to Austin.

In 1974, the Japanese quietly announced an effort funded by the Minister of International Trade Industry — about $250 million with talent drawn from competing Japanese semiconducting firms. They did not know what products were going to emerge; it was an investment to create new technology. There was clear understanding that they were not creating a cartel. The companies that contributed people would take the technology back to their own companies, and they would go to the marketplace in competition. The premium was speed — speed to convert new technology into marketable products.

The pathbreaking Japanese initiative did not get much attention in this country. But an interesting character in Minneapolis, Mr. William O. Norris, founder and chairman of Control Data Corporation, issued a few complaints because the United States did not have a mechanism to do the same thing. He was worried that Japan would short-circuit our lead in the vital information handling industry.

Four years later, the Japanese scored their great success in the international marketplace with a 64K memory chip and ceramic production techniques. Suddenly American business took note. The media played it up. Frankly, it was more often portrayed as unfair competition than as recognition of a unique approach.

In October 1981, with uncharacteristic fanfare, the Japanese announced a new program; initially almost $500 million, it now looks more like $700 million. It was designed to take world leadership in a revolutionary kind of computer featuring artificial intelligence. It was called "the fifth generation," and it would lead the computer industry, the fastest growing segment of the international economy with all the potential jobs and market opportunities.

That action lead Mr. Norris to issue Thunderbolts. American action was demanded. Senior executives of 15 computer and electronic corporations gathered in Orlando, Florida. Although they were diverse executives, they all had one thing in common: They were concerned about antitrust.

Out of that meeting came some basic decisions. One was to take a new approach — not a Japanese approach, a U.S. approach, but one that paralleled the demonstrated successful effort in Japan. The participating companies would pool resources and pool talent, with a clear understanding that although the technology emerging would be owned collectively, the individual corporations would go to the marketplace in competition with one another.

Teams were put together to define the nature of the research that would be undertaken. And the challenge was given to Control Data to structure the company now known as the Microelectronics and Computer Technology Corporation (MCC).

There are some unique features about the bylaws of this first-of-its-kind joint research venture. First, companies get in for a very small sum of money. It was originally $150,000, then $200,000 and now $250,000 to have their name on the letterhead. But the real cost comes from the fact that they are bound legally for three years to fund at least one of the major programs. If they want to leave at the end of three years, they can — with empty hands. The programs are designed for six to ten years in duration and these corporations will receive a return only if they wait until we have licensable technology to go to the marketplace.

The second feature gives the sponsoring corporations three years to use the new technology. The company that funds an area of research gets three years lead at no cost to do their own packaging, to manufacture products, and to penetrate the marketplace. After three years, MCC is free to license other companies; shareholder firms who did not invest in a research program get no advantage over firms that are not owners of MCC.

Not all the shareholders were enthralled with this provision nor were they eager to see a very aggressive licensing program. The latter is important to me because a portion of the ultimate revenues comes back to MCC. And that is what I have to offer to the talent whose ideas produced the technology in the first place.

There is another good feature. These 14 corporations must begin looking at new approaches at technology transfer, and any impedence to incorporate our newly created technology into this product line and get to the marketplace early will allow competitors to benefit from the money they put up into research.

The areas of research include advance computer architecture with four independent programs: artificial intelligence and expert systems, knowledge-based systems, database machines, and database management — all centering on the interface between humans and computers. We must remember that we are not designing computers for the current populace, some of whom may still be afraid of computers, but for those who are going to be coming out of the school system in the future having been exposed from the outset. I suspect "user friendly" will have substantially different meanings than it does today.

Another area is parallel processing. For the noncomputer buffs, every computer, no matter how fast it operates, functions sequentially, linearly, one step at a time. We must make the breakthrough to use computers in parallel, just as our brain functions, numerous steps at the same time. We are not talking about computers replacing the human mind. We are talking about the ability to store acquired human knowledge and to make it accessible and usable in ways that we have never before envisioned.

The other three programs are software technology, trying to find totally new approaches to engineering software, packaging integrated circuits, and computer-aided designs. Overall our objective is to accelerate the whole state of the art of being able to use computers more vigorous in solving both fundamental design problems and practical productivity problems of automating factories and offices.

Our commitment to excellence included the clear understanding from the outset that the corporate shareholders would nominate talent and that the chief executive officer of MCC would accept or reject that talent. If he did not find the requisite human resources, he would go outside to find it.

In planning the initial structure, I tried to focus not on what gets you into fast operations but on the problems needing solution out ahead, for the six- to ten-year duration of these programs.

History tells us that for new starts in industry and in government, you can assemble new talent; but after four or five years, much of that talent begins to flow into other areas. Some go back to parent companies and parent agencies. Others go off with their own ideas to take their shots at the American dream.

I was vitally concerned that we try to position this new research consortium where we could be top choice for the brightest talent being trained at the graduate level in the years out ahead. This was a prime reason why MCC chose Austin, Texas. The key decision drivers left little margin for those cities that competed. First, the scale of collaboration among state and local government, the academic community, and private industry exceeded what we saw anywhere else in the country.

I've been accused from time to time by officials of losing cities who were pursuing us that my having been a graduate of the University of Texas at Austin was in fact the real determinant that led us there. And to tell you the truth, it did play a role on one occasion, on March 18, 1983, in Chicago when the governors and large delegations from Ohio, Michigan, Minnesota, Colorado, Texas, and Georgia came to make their case why MCC should consider one of their cities. Fairly late in the schedule, Governor White of Texas came in with Chancellor Art Hansen from the Texas A&M System and Chancellor Don Walker from the University of Texas System. As they got started talking about working together I had to stop the proceedings to tell my colleagues from these nine different companies what an amazing event they were watching. In my days at the university those people wouldn't have spoken to each other much less cooperated, and here they were seriously commited to work together in collaboration and in collaboration with industry and state and local authorities.

Another critical factor that emerged from that session was about commitment to world-class education quality. A senior vice-president from one of our stockholder companies asked the chancellor of the University of Texas System if they were comfortable sitting eighth in the ranking of the national academy of science, or if there was still some momentum to become first. That implied criticism was apparently a surprise; six days later the chancellor called to tell us he had raised about $3 million and wanted to know who in the country he should hire? And that was the early harbinger of the second decision driver for Austin, amplifying the collaboration among state and local government, academia, and the private sector.

Texas' commitment to use resources of the university system and accelerate them by enormous private-sector gifts in the area of computer technology has been matched in this country only by the federal government. If you wonder why the great growth in high technology has occurred in the past in Silicon Valley and in Massachusetts, look closely at the scale of investment by the Defense Advanced Research Project Agency (DARPA) and the National Science Foundation at Stanford and Berkeley and at MIT and other Boston-area institutions. So it is the commitment for that same scale of investment independent of the federal government that was the second major element that convinced us that Austin was the place where we ought to create the technology of the future for the information-handling industry.

On the occasion of the press conference announcing our presence in Austin, I noted bluntly that our shareholder companies voiced something less than overwhelming enthusiasm about Austin. It was remote and hard to get to. None of us, however, counted on the tidal wave of publicity that would follow. Nowhere in the country are those left that consider Austin provincial or remote. There is a can-do spirit in Austin that in too many places of this country is not alive in the same measure. So after we combine technology venturing, investing in universities, and building intersector collaboration, what is really going to make the difference is that can-do spirit.

Let me now consider several critical issues and questions I am often asked. The first is antitrust.

In the decision to go forward with MCC, the decision was made to simply keep the Department of Justice fully apprised of the preparation of this new corporation. Bylaws were submitted on December 7, 1982. Three weeks later the assistant attorney general announced that the Department of Justice would not object to the creation of MCC but reserved the right and had the intent of examining each individual program both for any restraint of trade and in looking at a potential market cartel. On April 30, we filed very detailed research and development (R&D) agreements for each of the programs I described earlier. In the intervening months we have been engaged in a good-faith effort with the Department of Justice to try to understand the implications of MCC. Frankly, I think we have done a good job in educating a bright group of lawyers in the Department of Justice who initially thought we really knew what products were going to come out but just didn't want to tell them. They now understand the nature of long-range R&D and that one cannot foreordain what technology will emerge, much less what products might conceivably arise from that technology six to ten years from now. The Department of Justice attorneys have also come to understand that we must consider an international marketplace not just a domestic marketplace. We are currently submitting final documents for that long research period. (There are, by the way, many lawyers who are a good deal more properous these last nine months — we have done our part for that profession.)

We will hear from the Department of Justice in a few months. I feel confident they will decide that our programs are absolutely within the scope of our antitrust laws. I would hope we might see some definitive written declaration. However, I think the odds are that we will get a decision not to interpose objection to the existing technical programs and existing corporate memberships. This is all MCC needs, but it isn't what the country needs. The country demands more, a clear signal. I believe American industry requires legislation that allows economies of scale in research in order to build competitive advantage in the world marketplace. I get calls frequently from other industries that are considering creation of like kinds of R&D consortia. They want to know what it is like to deal with 14 different shareholders. But the overriding question is antitrust. Until we get a clear signal from Congress we are going to see other industries moving to create the collaborative research centers we need for pushing the frontiers of technology and securing American economic strength.

The next question is compensation. Is there anything especially innovative in the compensation of MCC scientists who do the research, for example, certain project rights? Unfortunately, there is no equity available because MCC is privately held. The initial salary scale can best be described as austere. But ultimately we came out with a high industry scale and a very attractive feature; commitment for yearly bonuses of up to 50 percent of salary based on performance against objective (and not tied to profitability down the way). Furthermore, we are working to get an IRS definition that will let us describe the eventual return for licensing as capital gain. Then we would have something significant to offer over the long term.

Regarding the potential participation of IBM I will be brief. In the early stages of putting together MCC, a number of corporations were invited to join that did not become members. Each had various reasons for making the decision not to join. In some cases, the concern was the extent of their collaboration abroad. In others, it was antitrust. IBM recently made the first public statement whereby they describe their concern about giving away some of their technology. A close reading of MCC bylaws would in fact reveal mechanisms to buy the technology from shareholder companies. The real question would be giving away their current lead. In one of the early parts of this whole process we agreed that MCC should only focus on long-term technology. If you could already envision a product from that technology, it should not be done at MCC. I think that is probably as far as I should go and allow IBM to make any other responses.

How does our budget compare with the Japanese budget for developing the fifth-generation computer? We have less; that I can assure you. A lot depends on how many shareholder companies we have. And remember we are also covering more areas of computer technology. The floor created for our efforts is about $500 million for ten years. The Japanese effort for fifth generation, focusing tightly on artificial intelligence hardware, might be somewhere between $400 and 500 million. What begins to surface is other investments, the Japanese haven't talked about their effort in software, but I am persuaded they have a major effort underway. We may be in for some unpleasant surprises by their accomplishments in software. Given the additional shareholders that are still in the talking stage, I suspect we will have on the order of $600–650 million over the decade. But the money is not the prime issue as long as we're in the ballpark. It's a question of assembling talent and giving them the environment in which they will do creative things.

But developing new technology may not be the long-term challenge here at MCC or to our corporations. The long-term challenge may be in transferring the technology we create to the production lines, and in doing it in the time frame that ensures competitive advantage in the marketplace. If you look back over history, our allies and economic competitors have been so much swifter at using technology and quality control in their manufacturing procedures, giving them substantial edge. Ultimately we will judge MCC's success on one fact that is easily discernible: whether its shareholders' products make profits in the marketplace. What we create in a laboratory environment is just preparation; it's not battle.

6

The Economic Environment

Robert Ortner

Before I joined the Reagan administration, my first tour of government duty, I used to think, naively, that Congress worked on a two-year time horizon, i.e., every two years its members geared up for reelection. I have learned this is wrong: They are *always* running for reelection. And since many congresspeople seem to believe that their constituents always want more spending programs, it may help to explain our huge budget deficit.

The administration has not yet convinced the Congress to recognize the seriousness of the deficit and its responsibility in reducing it. Bringing the deficit under control will remain a major goal of Mr. Reagan's second term.

In reviewing the U.S. economy, it might be helpful to discuss developments from two perspectives — the current situation and longer-term prospects and problems. These perspectives are quite different. The current picture is bright and clear. The long-term picture, as we say in Washington every summer, is hazy, hot, and humid.

RECESSION AND RECOVERY

First, the current situation. In the election campaign you heard many arguments over who was responsible for the last recession.

When Mr. Reagan was inaugurated in January 1981, inflation had been running at about a 13 percent rate (in some months of 1980, it was close to 20 percent) and the prime interest rate had just reached 21½ percent. The economy was already in trouble. In fact, two key industries, autos and housing (which are the most interest sensitive) — had been declining for two years.

It was said that this was the worst recession since the great depression. Actually, the 1981–82 downturn, as measured by declines in the broad indicators of activity, such as the gross national product (GNP), and industrial production and employment, and by the rise in unemployment, was close to average and shallower than the 1973–75 recession.

What made this recession seem so severe was that unemployment, which went up by a typical 3 percentage points, reached 10¾ percent, a postwar high. The problem was that the unemployment rate was 7½ percent when the recession started, which was also a postwar high for a business cycle peak. It was that high because the economy had not grown since early 1979, and there was a sharp, though short, recession in 1980 from which we did not fully recover.

The proximate cause of our economic stagnation was inflation; we let it get out of control. And the inflation was followed by record-high interest rates.

Another important factor, and an old problem in the U.S. economy, was our lack of capital spending. From the end of World War II to about 1973, real capital stock per worker grew at an average of close to 3 percent per year and productivity did almost as well; from 1975 to 1980, capital stock per worker increased only about 1 percent per year. And if we subtract pollution control equipment, the performance was worse. The result was that productivity stopped growing, and the economy followed suit.

Clearly, we do not invest enough. During the postwar period, business fixed investment in the United States has averaged about 10 percent of GNP, generally less than in Europe and much less than in Japan. One should not be surprised, therefore, to find that the U.S. record of productivity growth is one of the poorest. I do not only mean that we are behind our strongest competitors, Germany and Japan. We are also behind England, France, and Canada, i.e., behind every major industrial country.

Could the 1981–82 recession have been avoided? I do not think so — not after we lost control of inflation and interest rates. If the Federal Reserve had pumped up the money supply in 1981 and pushed down interest rates, the recession might have been delayed, but the resulting inflation and later collapse would have been much worse.

The recession ended in November 1982 and the economy has recovered very well. At the beginning of last year, the consensus seemed to be that the upturn would be weak. But the consensus was wrong. During the first year of this recovery, industrial production rebounded almost 16 percent compared with an average of a little more than 13 percent for the same period in the prior six recoveries.

The recession and recovery also appear typical in terms of patterns within the economy. Among the major sectors, homebuilding turned down and up first, as it usually does. Housing starts reached their trough in November 1981 at an annual rate of about 840,000 and rebounded to a level of nearly 2 million in August. Then consumer spending began to pick up. Its strength last year probably resulted from several years of deferred demand as well as the tax cuts and a normal cyclical turnaround. Sales of domestically produced cars, which languished at a rate of 5 million in the summer of 1982, are now running at about 8 million.

The economy also underwent a customary, if somewhat severe, inventory correction, which seems to have run its course. Capital spending ordinarily follows the rest of the economy and increased more strongly than usual, encouraged by the 1981 investment incentives.

NEAR-TERM OUTLOOK

Real Growth

From the end of 1982 to the fourth quarter of 1983, real GNP rose more than 6 percent, in line with the average of previous first-year recoveries. And in 1984, I think that growth will be in the area of 4 to 4½ percent, assuming that the Federal Reserve is serious about its monetary targets. And for the sake of inflation and of the long-term health of the economy, I hope they will stick to their targets.

That is not as ominous as it may sound, because these targets are not really restrictive. For example, this year's target growth rate for M2 was tentatively set at 6.5 percent to 9.5 percent. In 1981, the target was 6 to 9 percent and the actual increase was about 9½ percent. I refer to M2 because it contains transaction or near-transaction accounts not included in M1 and because over long periods of time it tracks the economy very well.

The point is that in 1981, with about 9 percent inflation, monetary policy was tight. Now, with slower inflation, the same targets will accommodate satisfactory real growth. However, with interest rates so high, and with a large treasury deficit, that growth path could be bumpy.

Inflation

Inflation has been a pleasant surprise. From those disastrous figures several years ago, the increase in the current price index (CPI) during 1983 was down to 3.8 percent. Of course, the slowdown was aided by the drop in oil prices and by stable food prices. When prices in both of these areas increased, inflation was higher in 1984, but probably not more than 5 percent. In the case of food prices, we contributed to our own trouble my taking land out of production under the PIK (Payment in Kind) program. And then, in accordance with Murphy's Law, we had the drought as well.

We have also had a healthy slowdown in wage increases. The Labor Department's Hourly Earnings Index rose less than 4 percent during 1983, whereas total compensation among nonfarm businesses rose 5 percent. In addition, productivity showed good cyclical pickup, with the result that unit labor costs increased little through 1984. The short-term outlook for inflation, therefore, is still very good.

Employment

The most important bottom line of the business expansion is the recovery in the job market. Since the end of 1982, total civilian employment rose by over 4 million and the unemployment rate dropped by almost 3 percentage points. The improvement in these indicators has been faster than usual.

LONG-TERM PROBLEMS

Thus, our current situation is better in nearly every respect than most people expected; yet there is some uneasiness, which, directly and indirectly, is related to the budget deficits. Because of high interest rates, homebuilding activity has already peaked and even slipped back. Our foreign trade is an even bigger problem. In fact, it is in shambles — devastated by the high dollar, which was pushed up by our high interest rates, high in part because of our huge budget deficits. These interest rates and budget deficits cloud our longer-term outlook.

Foreign Trade

In 1982, our merchandise trade deficit was $43 billion, with imports valued to include insurance and freight. In 1983, the shortfall reached nearly $70 billion. And in 1984 the deficit will rise to more than $100 billion, even if the dollar begins to fall.

The dollar is a large part of the problem, but it is not the whole problem. Our strong recovery is drawing in more imports. At the same time, our exports have been stagnant, partly because of the high dollar, and partly because of sluggish economies abroad. Today, in effect, the United States is the world's economic locomotive.

Our exports also have been hurt by debt problems of developing countries. Over the past two years, U.S. exports to Mexico dropped by $9 billion, about the same amount as the decline in exports to Europe. U.S. exports to Mexico appear to have stabilized, but the debts of the less developed countries, especially in Mexico and Brazil, will be problems for the international financial community for many years to come. As these obligations are denominated in dollars, their burdens are made worse by the strong dollar and high interest rates in the United States.

What should we do about our trade problems? One approach that we have begun to take, unfortunately, is protection. Another approach, fortunately still only a proposal, is an industrial policy of targeting individual industries. In my opinion, both approaches are wrong.

Protectionism

We are already protecting a number of industries, including autos, steel, and textiles, and the House has passed a bill providing further aid for our auto industry in the form of local content requirements. International retaliation to these moves will hurt everyone. What is not understood is that even without retaliation, we hurt ourselves with restrictions. Limiting imports adds to our inflation not only because many of our costs and prices are relatively high, but also because our domestic producers are encouraged to raise costs and prices still more.

Private estimates indicate, for example, that the local content legislation, if it were to pass the Senate and become law, would raise car prices 10 percent initially, cost American consumers $18 billion per year and ultimately result in lower employment in the economy. The irony is that, as a *New York Times* editorial pointed out, the government, by maintaining import barriers, is "forcing Americans earning $10 or $15 an hour to guarantee the jobs of auto workers earning $26."[1] In effect, the auto industry is asking the government to play not Robin Hood, but the Sheriff of Nottingham.

[1] *New York Times*, September 10, 1983, page 22.

Industrial Policy Initiatives

Mainly because of foreign trade pressures, a number of proposals have been made under the general title of "Industrial Policy" to promote and subsidize certain industries. In one form or other, an industrial policy solution to our trade problems is being pushed by the affected companies, which represent older, "smoke-stack" industries, and by their unions. The idea also has found a sympathetic ear and enthusiastic support among legislators with traditional "do something" habits. This supposed solution to pressing problems should also be rejected.

Proponents of industrial policy base their argument on three premises, all of which are wrong. First, that U.S. manufacturing has been eroded; second, that industrial policies have been an overriding success abroad; and third, that government can make better business decisions than business people can.

On the first point, to paraphrase Mark Twain, reports of the demise of U.S. industry are greatly exaggerated. A recent study by David Lawrence, published by Brookings Institution, shows that industrial production grew faster in the United States between 1973 and 1980 than in Germany, France, or England (although not as fast as in Japan) and that manufacturing employment grew faster than in any other industrial country, including Japan. A Commerce Department analysis indicates, moreover, that employment losses in U.S. "smoke-stack" industries will be more than made up by gains in the services and "high-tech" areas (although not necessarily in the same geographic locations as the losses).

Reports of foreign successes in industrial policy ventures also are greatly exaggerated. Among the apparent favorable outcomes are Japanese semiconductors, French energy development, and the European Airbus. The Japanese steel industry is often cited as an example of success, but the industry has not been particularly profitable and now, with growing worldwide competition, some capacity has been shut down. Other clear failures are German attempts to develop a computer industry, Japan's investment in shipbuilding, and the Anglo-French Concorde. It is also hazardous to assume that what did work earlier in Europe or Japan would work here, or even there today.

The third premise, that government decision making is better than private, is the weakest of the three. Politicians, or their appointed "experts," are not as familiar with any industry or its problems as those dependent on it for their livelihood. Decisions by politicians will reflect political rather than economic criteria.

Even if these industrial policy programs had been an unqualified success abroad, it does not follow that they should be pursued in the United States. We already have a much more efficient capital allocation system than exists in other countries. It is known as Wall Street. Advocates of an industrial policy claim that new companies, as well as older ones, are unable to raise capital privately. That is simply not true. In 1983, new issues by high-technology companies totaled $13 billion. And private placements raised billions more. Eventually, the marketplace will separate the successful from unsuccessful companies. What it will not do is continue to pour money down the drain of failures as government policies, under lobbying pressures, frequently do.

Many of our apparent industrial problems are really macro in nature, a result of erratic fiscal policies, large budget deficits, and high interest rates. Targeting industries for subsidies, loans, and protection may help some of them but will surely worsen the economic pressures on others. The efforts are likely to be self-defeating. So my advice to Congress is, as my friend and former colleague Murray Weidenbaum once said, "Don't just do something, stand there."

Of course, introducing legislation either for protection or for subsidies to targeted industries is easy to do politically, even if it is the wrong measure economically. The right measure is to cut the budget deficit, certainly not increase it.

Budget Deficits

The deficit in fiscal 1983 was $195.4 billion and the latest estimate for 1984 is $184 billion. The deficits will not disappear and cannot be ignored. Each year of a $180 billion deficit will add about $18 billion, at today's interest rates, to subsequent year's interest expense. And these ongoing deficits, in turn, serve to keep interest rates high. Therefore, the problem feeds on itself and will tend to get worse. Time is not automatically on our side.

There are also several misconceptions about these deficits that I am sure you hear often:

- Deficits do not raise interest rates.
- Deficits are not a problem because economic growth will increase receipts and cure them.
- Deficits will slow or stop the recovery by crowding out private borrowers.

Obviously, these positions are not argued by the same people, as the second and third are contradictory.

Given credit demands in the private sector and the Federal Reserve's targeted growth of money and credit, heavy treasury borrowing maintains pressure on the credit markets and does hold up interest rates. Wall Street seems to understand this very well, even if there is some confusion in Washington.

With regard to the second fallacy — that economic growth will balance the budget — of course, growth will increase receipts, but beginning in 1985 our tax rates will be indexed, whereas spending will continue to rise both from inflation and growing defense outlays. Without new spending cuts and possibly even some tax increases, the deficit will not shrink.

The first two positions — deficits do not raise rates and the deficits will be self-correcting — are actually dangerous because they imply that we do not have to do anything about the budget.

The third misconception — that the deficit will strangle the economy and kill the recovery — is simply wrong. A big deficit is highly stimulative. It is not restrictive. Monetary policy is what controls economic growth and without this control we would be back quickly into runaway inflation.

What the deficit will do is distort the composition of growth. Interest-sensitive sectors will be depressed, especially foreign trade, housing and to some extent autos. There has been a long-term trend away from smoke-stack industries toward service and high-technology industries. Because our older industries are more heavily engaged in and vulnerable to foreign competition, the large budget deficits, high interest rates, and strong dollar have accelerated this trend. And, eventually, if the budget deficits are not reduced, they will lead to renewed inflation, still higher interest rates and another severe recession.

HARD DECISIONS LIE AHEAD

How did we get into this mess? The source of the problem is actually simple; solving it will be difficult. The seeds of this deficit were sown by the buildup of social spending throughout the 1960s and 1970s. (Thus, the problem is political.) Since 1960, the cost of social programs rose from about 4 percent of GNP to 12 percent — three times as fast as our ability to pay for it. At the same time, defense spending fell from 9 percent of GNP to under 6 percent. During the same period, the effective total tax rate on taxable income crept persistently upward, from 15 percent in 1960 to 24 percent in 1981. After three rounds of income tax cuts, the rate is still a high 22.5 percent and is beginning to creep upward again. If the Congress and some past administrations had been content with increasing social spending only twice as fast as GNP, we would have virtually no budget problem today.

Obviously, much work is ahead on the budget, but it is not likely that Congress will cut spending, especially for the social programs, or raise taxes (which the administration does not want to do) in an election year. A genuine long-term solution will require that we change our well-entrenched spending and taxing habits — habits that have always been detrimental to investment and to inflation.

If a solution must include more taxes, I would prefer taxes on spending so that saving and investment incentives will remain intact. We must stimulate research, innovation, capital spending, and productivity if we are to have a healthy economy. Growth in productivity is the source of increased real wealth and improving standards of living and, therefore, underlies the true economic welfare of the nation.

7

Strategic Implications of the Changing Economy: Business/Industry Perspective

Francis P. Cotter

What are the strategic implications of a changing economy? To do justice to this subject, it is important to first look at how far we have progressed. My perspective is from business and industry.

Past societies solved their industrial problems over periods bounded by centuries. More modern societies have evolved solutions over intervals measured at most in decades. But American industry today is challenged to shift from a domestic to a global economic orientation in a time frame measured in months.

This follows the late 1960s and early 1970s, when some national figures were speaking about limits of growth. The zero-growth advocates took a line from Ogden Nash, who said, "Progress might have been all right once, but it has gone on too long."

Then it was said that we suffered from a malaise, which I think is something you catch from swallowing limits of growth too much. Today, we are back to reality and are talking about industrialization, of sunset industries, falling productivity, and a postindustrial society. And the proposed remedies of Industrial Policy are a government board to pick industrial winners and losers and a bank to fund the fortunate choices.

Not having the credibility of an economist or the wisdom of a Phil Donahue, I may not have such a profound approach; but I have three suggestions for the future.

First, as I view the laws surrounding industrial competition, the antitrust laws, I believe we must recognize the importance of *industrial efficiency* as a key element in the societal needs that frame our laws. We no longer can afford blindly to break up an AT&T or any other organization that efficiently serves our society. We can no longer afford to introduce high technology, as the nuclear industry has tried to do, into hundreds of individual entities that lack an interconnection in the name of antimonopoly.

Second, we have to introduce a new realism into our federal export policies so that the new technologies we develop can enter the world's markets. We need to reap the economic benefits of exports to afford larger research and development expenditures for better product development.

Third, instead of sitting back and letting government take the lead, industry should establish some institution that monitors the changing world economy, the changing sciences, and the changing technologies. I have in mind something like a National Science Foundation for American Industry, an interindustry organization that would try to define what is happening in the global marketplace.

Let me explain these three ideas.

The rules of economic competition in America are prescribed by a host of laws; this is the most litigious society history has ever known. The very ability to enter into competition means that one must first go through a government toll booth. Anyone connected with potential antitrust situations knows that government approval is harder to obtain than being found innocent of heresy during the Inquisition. Government also runs the toll booths leading outside of our nation through its trade regulations or its use of export controls to attain foreign policy objectives.

Our antitrust laws were enacted at roughly the turn of the century. The America of that period was emerging from an agrarian society into an industrial one. The Sherman and Clayton Antitrust Acts addressed social, not economic, needs. They addressed price competition questions and the preservation of small business and focused on problems such as the railroads exploiting the farmers. They could hardly have anticipated a global economy of satellite communications, commercial aviation, or massive electrification. Looking at my industry, Edison built his first centralized electric power plant, the Pearl Street Station, two years after the passage of the Sherman Act.

In the ensuing years, a vast body of legal precedent went into book after book, expanding the scope and reach of antitrust. Whenever a business was embroiled in an antitrust matter, the instinct was to settle — let's get rid of the problem. This allowed almost every precedent to increase the power and reach of the government.

But in a global economic system, these antitrust laws are outmoded. If there were totally free trade in automobiles, if quotas were not permitted and price alone were our only concern, it would make little difference if General Motors and Ford combined. The rest of the world would pour their products into the U.S. markets and prices would be stable.

Similarly, I think our interests in this country would be better served if our steel companies could combine for the purpose of building one steel mill and have it be the most efficient in the world. From such a demonstration plant badly needed technologies could arise.

In my view, our antitrust policies will force us into protectionism. Quotas force prices up, and we get to see the "Washington Rule or Unintended Consequences" working at its best. Laws enacted to force prices down inevitably raise them.

Generally, firms with large market shares have been considered bad *per se* — politically, if not legally. In a global economy market share should be measured across continents, not states. In fairness, I must say that the Department of Justice is starting to take a fresh look at things and I wish them Godspeed.

Another example, something closer to my experience, demonstrates the problems of splintered industries like the nuclear industry. In the case of the nuclear navy, where only Westinghouse reactors are used, they are produced and sold to a single entity, the U.S. Navy. When we took the same technology to the civilian market, we have multiple entities at multiple levels — some four manufacturers supplying components to an array of architect-engineers and their subcontractors to be assembled by scores of utilities, all under the smooth guiding hand of the Nuclear Regulatory Commission, who some call the "A Team." In the electrical equipment industry, one sells to 234 investor-owned utilities, 1,033 rural electric utilities, and 2,195 municipal bodies.

The chairman of the Nuclear Regulatory Commission, Nunzio J. Palladino, in looking at this troubled industry, said recently, "I think our major problems are with the smaller utilities who didn't have an appreciation for what they were getting into and didn't develop within their organizations a spirit of getting the thing built properly."

We are starting to hear suggestions that the nation should have only big electric generating companies and today's utilities would mainly distribute power. I can't judge the merits of this proposal, but how would you like to guide that change through Congress?

My second point is the current U.S. export policy.

The effectiveness of the competition American industries now face in world trade is, in large part, a testimonial to the success of our industrial policy. It was our success with the Marshall Plan that rebuilt foreign economies after World War II. It was the expansion of our economy that gave these foreign industries the markets for their goods. It was the American educational system that provided and continues to provide many of the scientists and engineers to design, run, and expand foreign industries. And it was American discoveries, know-how, and technology that provided the spark to the Atomic Age, the Space Age, and the Information Age.

Today, with only minor exceptions, it is a fallacy to think we have sole control of world markets, products, or technological development. Libraries are full of the expanding knowledge of technological wonders developed around the world, and our patent office is crowded with foreign visitors. Over one-fifth of all graduates of American universities with degrees in science and engineering are foreign nationals who form the majority of our graduate students and research fellows.

Perhaps nowhere is this fallacy of continued American technological domination more pervasive than in the field of civilian nuclear power. Just consider the facts. The secrets of the atom were unraveled largely by scientists who were driven to our shores during the rising tide of fascism in Europe. If we had a monopoly, it has long since disappeared. Of the 300 nuclear power stations in the world, we have only 80 in the United States. But we constantly hear that if we export nuclear power stations it will help spread nuclear weapons. Of the five nations that are now nuclear weapons states, no weapons were developed from their commercial nuclear power programs. In fact, all developed weapons before they developed nuclear power; China, a nuclear weapons state, still does not have nuclear power.

No nation would realistically use civilian nuclear power to develop weapons. There are much easier, cheaper, and more readily available ways to do it. If commercial nuclear power is a viable means for weapons production, why do the Soviets allow their satellite states to have such power plants? Yet they all have them and the Soviets plan 50 more stations in satellite countries.

U.S. nuclear export policy is driven by the theory that since the used fuel of the power station contains plutonium, it could be separated to make a weapon. But power-reactor plutonium is almost unusable for weapons. But even if it were usable, how can the United States with only 30 percent of the world market in nuclear fuel control events through onerous domestic laws? It can't.

Most recently, the absurdity of our policy was exemplified by the inability of American nuclear equipment manufacturers to sell to China, already a nuclear weapons state. But efforts are being made at the presidential level to reach an agreement with China.

Thirty years ago President Eisenhower addressed this problem of nuclear nonproliferation by bringing about a system of multinational nuclear controls, which works well indeed. The International Atomic Energy Agency does a fine job of inspecting plants in over 100 nations. But we couldn't leave well enough alone.

We also have serious problems with our ability to finance exports. The government agency that was designed to assist us, the Export-Import Bank, budgeted $3.8 billion of lending authority for this year. Until now, they have committed only $17 million. When you look at our balance of trade deficit, this is not a sterling performance.

Last, we have been hearing much about Industrial Policy. There are legitimate concerns regarding the viability of our basic industries in light of world competition. The healthy part of all this is the lively debate led by those who, only yesterday, were speaking about zero growth to save our environment and to save allegedly vanishing resources. They have changed their tune.

Clearly American industry, the American worker, and American students are greatly affected by the winds of change sweeping across the continents. Information flows around the globe at the speed of light. Knowledge and know-how not only can move in jet planes; but with computers, satellites, and telecommunications, it is now becoming possible to be in two places at the same time.

But of all that has been written about these changes, for all the discourse and discussion, how much do we really know?

The American worker is neither inherently lazy nor unproductive. What is his or her role in tomorrow's society?

We are not Japan, and our need for capital formation cannot be solved the Japanese way. What is our alternative?

We are not France, and we cannot nationalize our basic information — even if our industries must compete with the French government. How can we better aid our industries in such biased competitions?

We are not a socialist economy committed to a planned economic system whose efficiency is so poor that it must steal secrets from others to try to catch up.

I can raise other questions — they go on and on. I do not know the answers. But they are important questions. We cannot afford to ignore them.

Having spent most of my adult life in Washington, I know the answers will not be found by a board of labor leaders, bureaucrats, consumerists, environmentalists, or industrialists. What we need are some thoughtful analyses, based on solid factfinding that results in realistic solutions. What we need is something akin to a National Academy of Science, an industry-sponsored Academy of American Industry chartered to conduct continuing studies in the development, growth, and efficiency of American industry in a global economy.

The problem is ours; and until we start making it ours, we are not likely to be happy with the solutions others will give us. We in American industry must marshal our many resources and look to our own experience and skills. We can provide the American public and the government with the facts, the ideas, and the commitment to maintain our preeminence in today's and tomorrow's world economy.

8

Financial and Investment Perspectives on Technology Venturing: A Private-Sector View

Cordell W. Hull

Technology venturing involves not just high technology but also the development and construction of large projects. Technology venturing is especially important when viewed in the context of natural resources, energy, and basic industries, which to a large extent is the arena in which my company, Bechtel, operates. Continued technology venturing in these areas will be very important for the sustained economic growth of this country and indeed for the rest of the world. Such growth is necessary and desirable.

In developing the financial and investment issues currently being faced by technology venturing, I would first like to describe how technology venturing evolved in these areas. Back in the early 1960s, Bechtel built a 600⁺-megawatt oil-fired power plant for a U.S. utility client at a cost of slightly over $60 million in just two and one-half years.

Recall those days from the standpoint of new endeavors:

- A rather conventional technology was involved.
- Inflation at the time was just under 2 percent.
- The utility financed those plants with 20- to 20-year bonds at around 4 percent fixed-rate money.
- Equity was sold by the utility at 30 percent above book value.
- Capital markets were strong, predictable, and healthy.
- The dollar was stable and reasonably priced.
- Regulations did not unreasonably impinge on construction schedules.
- There were no intervenors.

Today, a similar project might be coal instead of oil fired and would take upward to seven years to complete. It would cost over $500 million and licensing delays would be a new major risk. Interest and inflation during construction could run as high as 50 percent of the final costs of the project.

Moreover, the utility could expect:

- Great difficulty in raising long-term fixed-rate money with costs running as high as 14 percent per year in a thin and highly priced market.
- Sales of equity stock to finance the plant at book value or below, thereby diluting the interests of existing owners.
- Immense paper work and costs in dealing with regulatory compliance matters.
- More coal gas stack scrubbers, new burning and pulverizing techniques, and other environmental treatment aspects.

Internationally, the trends are quite similar. For example, Bechtel built the 130,000 barrel per day refinery in southern Belgium for a major U.S. multinational oil company in the late 1960s at a cost of $100 million in just two years. The client, Standard Oil of California, was the sole sponsor of this project and financed it directly out of its own corporate resources. Today, comparable projects cost up to $1 billion and take twice as long to build. For example, the 14,000 barrel per day Gas-to-Gasoline project in New Zealand will cost $1.5 billion and take about five years to build. This project, which is the first commercial-scale synthetic fuels venture of its kind, is sponsored by a joint venture between Mobil Oil, another oil major, and the government of New Zealand. Eighty percent of the project's costs is being debt financed in international markets with a package of government-subsidized export credits and international commercial bank loans on a project financing basis; that is, on the basis of the project's economics and cash flow as opposed to strictly the creditworthiness of Mobil or the Crown. I could cite many related examples from projects in other industries.

In this environment, financial and investment matters have become considerably more complex. As we moved through the 1960s and into the 1970s, capital and financial markets changed for all major technological projects:

- Inflation soared up to almost 15 percent and with it interest rates rising to over 20 percent.
- Economic cycles became shorter, more intense, and oscillated with increasing amplitudes.

- Because of regulatory constraints, larger project sizes and other factors, project development time lengthened to 4, 8, and even 12 years.
- Project costs became virtually unpredictable when interest and inflation costs were compounded over these longer periods — a $1 billion dollar project was not unusual and some projects cost over $10 billion.
- At the same time, project developers had to cope with escalating regulatory uncertainties and ever-mounting intervention from groups and individuals representing special interests.
- Long-term, low-cost fixed-rate money dried up.
- The dollar ceased to be the bedrock of international foreign currency stability as the Bretton Woods system crashed around us.
- Borrowers throughout the world overextended themselves in the flush capital markets with short-term, floating interest rate loans creating havoc.
- Major project undertakings were increasingly sponsored by large consortia of owners and financed in increasingly complex and unconventional ways.

It is little wonder then that American industry postponed essential investment in plant and equipment and often focused on short-term and quick-fix approaches.

Then in the early 1980s we saw the culmination of these excesses as the world economy collapsed around us after one final orgy of inflation and financial excesses. Now, as we move forward into economic recovery, what are the implications for future technological projects and the issues that bear thereupon?

As to the world economy itself, I am optimistic. Certainly, the recovery in 1983 in this country exceeded most expectations by a significant margin.

We now see the increasing impact of our recovery spread throughout the world as its effects cascade through the natural resource, energy and heavy industrial sectors creating demand for new technology venturing. Furthermore, we are on the verge of economic changes potentially as dramatic as anything since the industrial revolution commenced. This should bring with it economic growth and opportunities unparalleled by anything in our lifetimes.

We have no real choice but a massive rebuilding of our basic industries, supporting the upgrade of industrial and social infrastructure to the most competitive plant in the world. We must also restructure our management techniques, institutional approaches, and business methodologies to compete efficiently in a world market where our massive natural resources, people, technology, raw materials, and a great agricultural base, give us a natural competitive edge.

The cost of this, however, will not be cheap. It has been estimated by Data Resources that U.S. corporations will require more than $3 trillion in capital in the nex ten years alone to modernize and rebuild plant and equipment. Rebuilding long-neglected U.S. infrastructure so necessary to serve new industry and technology venturing will require trillions of dollars more. Moreover, there are serious issues that must be resolved with respect to forward technology venturing. This can be illustrated in several areas where Bechtel is involved.

First, Energy Transportation Systems Inc. is planning to develop a coal slurry pipeline with which Bechtel has been associated for over 15 years. This is technology venturing on a grand scale. This multibillion dollar project is expected to transport up to 30 million tons of coal per year in slurry form over about 1,400 miles from the Powder River Basin in Wyoming to electric utilities in the southcentral United States. However, after spending over $100 million in development efforts, this project still has not been built in spite of its clear advantages and economics. Although the reasons for this lengthy lead time are complex and relate to regulatory, marketing, financial, environmental, competitive, and other matters, it clearly demonstrates that high front-end risks, long lead times, and the interests of a wide variety of parties both in the public and private sectors are intrinsically involved in the development of modern big technological projects.

Second, technology venturing today may involve billions of dollars in risk capital. The building of a 1,000-megawatt nuclear plant will cost $3 billion, a single 50,000 barrel per day oil shale plant will cost $6 billion. This will mean heightened vulnerability to sudden market downturns or changes.

We are all familiar with our synfuel odyssey, where the complexities of dealing with a government agency for funding or guarantees, huge project development costs, long development periods, uncertain future energy prices and demand, government controls, complex environmental factors, large multipartied development consortia, and a host of other matters have converged to stall the development of technologies that may be vitally needed for our country's future economic health and national security.

One synfuel project that has moved forward is an integrated combined cycle coal gasification plant, located in Daggett, California, which will produce about 100 megawatts of electric power using gasified coal as fuel and environmentally sound processes. This $300 million project is being undertaken by a consortium that includes Texaco, Southern California Edison, Bechtel, General Electric, and several associations representing the electric utility industry, including the Electric Power Research Institute. In so doing, interests representing technology, equipment, engineering, and energy have pooled together their resources and are taking considerable risks to develop an advanced technology that will be of mutual benefit to all.

Another important participant in this project is the Synthetic Fuels Corporation (SFC), which is backstopping it with $120 million in price guarantees to cover possible cash shortfalls during the first five years of the project's operation. This was the first such commitment of financial support made by the SFC.

The state of the private nuclear industry in this country represents a true national tragedy. Billions of dollars of plant investment are going down the drain as partially completed plants are scrubbed in state after state. This is a technology venture failure on a grand scale and one in which the blame must be allocated among a wide variety of involved parties, including the utility sponsors themselves, federal government regulators, engineer/constructors, intervenors, our judicial and related institutional structure, environmentalists, material and equipment suppliers, state public utility commissions, short-viewed politicians, and others. Yet, other countries have found ways to adapt their systems, institutions and policies to develop this technology economically so that it may enhance their lives and competitive industrial posture.

To state the obvious, a prerequisite for future technology venturing will be creating a favorable economic and financial environment. This will require:

1. Healthy capital markets.
2. Greater monetary stability.
3. Prolonged low inflation.
4. Reasonable interest rates; real rates relative to inflation are still excessive.
5. Policies to induce savings to support capital investment.
6. Stronger and more stable economic growth.

Whether these conditions can be achieved, and to what extent, is largely a question of political will and understanding. There is much to encourage us, as there is growing evidence that leaders of all mainstream political persuasions are increasingly directing their programs toward achieving these aims.

However, there are equally important requirements in the financial and economic area that have yet to be met:

1. A reasonably priced dollar and stable and predictable foreign exchange markets. It is now difficult to state whether our basic industries are really so uncompetitive against foreign products or whether the problem is merely a macroeconomic phenomenon created by overvalued dollar. I suggest that no small measure of this problem is treated by the overvaluation of the dollar by perhaps as much as 30 to 50 percent.

2. Recognition of the international dimensions of our problems. This is an aspect that we as a nation are only beginning to grasp. We can no longer look upon the domestic economy, our government and business policies, or our industries as anything but an integral part of a bigger international mosaic. Economic, financial, and business policies must be formulated with world markets in mind. Trade now accounts for almost one-quarter of our gross national product (GNP). Yet, fewer than 1 percent of U.S. companies account for 80 percent of our exports. In spite of the growing importance of trade to our economy, our government policies often are still made to meet strictly domestic concerns and fail to consider possible implications for our international competitive posture.

3. Development of some program for dealing with the $800 billion plus debt of the less developed countries. A program is needed to assure that the integrity of our banking system will be maintained and funds will be permitted to flow from that source on the ever increasing scale that is needed for our own development needs. The American financial system must provide the capital so necessary to support trade with the developing countries, enabling these countries to continue their own development programs.

4. A lower and more stable interest rate structure. We can no longer sweep under the rug the devastating impact of our uncontrollable budget deficits on our interest rate structure, the inflated value of the dollar, and the national psyche.

5. Abundant and reasonably priced energy supplies. These are essential to maintain a healthy growing competitive economy. Electricity is the backbone of industrial growth and decent living standards. We are on the verge of a national emergency in this area in the 1990s if the development of new electrical generating capacity is not started soon.

6. Changed economies of scale. We can no longer simply look at economies of scale in the traditional engineering sense. Now there are commercial, regulatory, financial, managerial, and other issues that bear upon the economics of a venture. The sheer scope of the undertaking may be too complex for the human mind to cope with in a reasonable and rational manner. The risk of failure of such a large-scale project may be far too onerous for private developers to assume regardless of the ultimate potential economies of scale.

7. Further refinement of techniques for risk sharing in project development. Technology venturing today so often depends on the financial and credit support of a number of the participants. For example, small hydro, geothermal, cogeneration, and other alternative energy technology ventures are currently being developed in a variety of creative ways with the support in the form of contractual assurances or guarantees from fuel suppliers, offtakers of power and steam output, and even engineer/contractors. In some cases, financial institutions have stepped in and assumed certain financial and commercial risks. Mechanisms for creating government safety nets may also be required in some instances, since longer-term economic stability, regulation and other government policy actions often directly affect the viability of such undertakings.

8. More flexible commercial and funding arrangements. We no longer live in a stable predictable world and it is naive to expect one in the near future. Who is to say with any degree of certainty that the era of oil price escalations, inflation bouts, and interest rate shocks is over? Political disruptions such as that incurred in Argentina in 1982 are increasing with some greater frequency and are virtually impossible to predict. Sufficient flexibility must therefore be built into commercial and financing arrangements surrounding technological undertakings to allow for inevitable unexpected volatilities that will occur and bear upon the economic viability of the undertaking.

9. Continuous innovation in commercial and business structures. New financing techniques, new government ventures, and new participants, including even contractors, manufacturers, and foreign venture partners, will be involved in projects in this country. Already, export credits from abroad are being used to finance factories here. In addition, we will be seeing new relationships between management, labor, and government that will be an important element in the development of many new technology ventures.

Large-scale technology venturing has become a more complex process, involving the melding of a wide variety of interests, concerns, and factors.

Fortunately, we have begun to have an increasingly better grasp of the necessary elements and conditions for the development and implementation of such ventures and are thus better able to plan and develop projects with contingencies in mind. Admittedly, we face an important challenge ahead.

Newer commercial arrangements are being formulated for these projects and we are seeing innovative risk allocation. What parties should be invited to join the undertaking, and how should risk be allocated among them? What should be the long-term contractual arrangements? What should be the type of conditions that permit flexibility in those arrangements? Increasingly, we are seeing foreign enterprises come into this country and participate in new ventures; this is going to be an increasing phenomenon, and one we should welcome.

We must provide in these projects significant flexibility for unexpected volatilities, be they economic, political, or competitive — anything that could affect the total integrity of the projects. We do not live in a stable, predictable world. There is no reason to believe that the world is going to become any more stable or predictable in the future. We can no longer look at projects just from a technological viewpoint. Engineers can prove there is a technological motivation for doing something. But if you look at the commercial, financial, political, and social practicalities, you may find that they dictate a different answer regarding economic viability. But public and private sectors are increasingly coming to grips with all these factors, and there is reason for remaining optimistic for new national projects in technology venturing.

9

Industrial Policy versus Creative Management: The Search for Economic Direction

Robert Lawrence Kuhn[1]

"Industrial Policy," the new call among political partisans, tickles our ears in an election year; it is, we are told, the national economic panacea for international competitive sickness. IP, to those on the in, would direct and control from Washington the thrust and focus of American industry. IPers believe that the free market system is no longer efficient and that the government must intervene to support business and prop up jobs. Coined by intellectuals and caught by politicians, IP is a symptom of economic illness and political fever.

One cannot deny the appeal to industries suffering decline and workers without work. Nor can one negate the fact that in a tightly wired world foreign governments can shift the commercial balance of power by giving home-grown companies unfair advantage. Thus, IP sparks the hope that federal funds might aid outmoded and out-priced companies in regaining former glory.

But numerous industries will vie for the golden tap. Which to promote and which to protect? Which to ignore and which to forget? When the government picks "winners," it must, by the same decision, also pick "losers." To sustain one, we must shun another. An increase of jobs here must result in a decrease of jobs there. If automobiles are chosen, why should textiles be condemned? Who is to decide that employment in the northcentral region should

[1]For a comprehensive description of top-performing medium-sized companies, see *To Flourish among Giants: Creative Management for Mid-Sized Firms* by Robert Lawrence Kuhn (New York: Wiley, 1985). Ten creative strategies characterize the best firms, each portrayed with numerous real-life case examples.

go up while employment in the southeast should go down? One conjures up tortuous visions of procedural miasma, politicking, and lobbying of unprecedented magnitude. Resources, we have come to learn, are not unlimited; available subsidy is only finite. (What, by the way, happens to IP when favorite industries do not make the federal hit parade?)

Socialism, it is said, is a wonderful concept; the dream of economic equality and financial fairness is utopian. The only problem, of course, is that it just doesn't work. Theoretical idealism breaks up quickly against the rocks of pragmatic realism. Human beings function best when they are controlled least, when they prosper in proportion to personal initiative and self-driven intensity.

American business is still burdened by archaic regulations codified two generations ago. There were right and rigorous reasons then. We were fast becoming, in those heady days, the world's premier industrial power; our growth was unimpeded, domestic markets were burgeoning and foreign markets beckoning. Industries and industrialists became intoxicated with their new-found powers, and consumers and workers, at the mercy of these mammoths, needed protection. Yet times shift and paths twist. What worked then, won't work now.

Is passivity the answer? Is public policy perfect? Should national debate go quiescent? The status quo be bronzed? By no means. What American industry needs is simple: Not more control by government but more confidence in management. Not centralized planning by bureaucrats but aggressive leadership from businessmen. Not Industrial Policy but Creative Management. Less macro, and more micro.

American industry must be freed from constraints, not encumbered with more. American business must be invigorated, not suffocated. The mold for forging the future? Independent management, not centralized command. (For every rule, there is an exception. In certain areas of the economy, especially in high-risk advanced technology, individual companies cannot afford to invest and America cannot afford to abdicate. Supercomputers, for one, have huge development costs and uncertain commercial revenues. Yet the United States must not lose world leadership, certainly not by default. Here is fertile substrate for an IP.)

The ends of IP are desirable, its the means that are questionable. It is not sufficient to deny IP for American business. To critique is always easier than to construct. IP will not work. What will? It is one thing to describe the illness, quite another to prescribe the remedy. Alternatives proffered usually stress macroeconomic manipulation, like looser money, tighter budgets, and the like. Yet something is missing. We've heard all this before.

One might believe by reading erudite arguments and counterarguments that industrial revival in America is linked to some "new economic policy," whether monetarist and supply side on the one hand or increased taxes and government spending on the other. A cardinal mistake here — and it permeates contemporary thought — is the notion that economic solutions to industrial problems will yield business success and competitive advantage. Macroeconomics surely has its place, but not the whole place. Macroeconomics is vital in defining and modulating the pace and proportions of the economy, but it is deficient in securing and prospering individual firms. It's like trying to coach a basketball team by determining the theoretically proper mix of heights, weights, and talents of players without ever teaching any of them how to dribble, pass, or shoot.

Economists dominate economic thinking. Logical, at least at first. But economists, when one thinks about them, don't run companies. They don't manage budgets and don't direct staffs. They never formulate corporate strategies and have no experience with corporate structures. "P&L," "personnel," and "product positioning" are terms they do not use. Meeting payrolls is something they do not do. Making enterprises work is responsibility they do not have.

Yet enterprises, for-profit businesses and not-for-profit institutions, are the components of the economy. Like cells in a body, they *are* the economy; and to treat the economy only by macroeconomics is to treat an epidemic only by epidemiology. Building businesses in the former, like curing people in the latter, must be addressed. To leave the economy solely in the hands of economists is to leave the sick solely in the hands of statisticians.

We must listen to the gross national product. We must hear the rhythms of small businessmen, middle managers, and corporate executives. We must feel the beat of individual needs, wants, and desires. The world works because some have vision and brillance, with the tenacity and temerity to produce and provide. Business, to me, is the economic synthesis of human knowledge, the molding of value and substance out of concept and form. It is the modern human analog of the original Genesis creation when the heavens and earth were formed out of chaos and void.

Creative and innovative management is what America needs, and government policy should be directed toward building it. But this is not a topic of macroeconomics; one does not study it in doctoral programs; there is little research, no Nobel Prizes, and minor media. It is local not global, micro not macro.

Yet the stakes are big not small: Creative and innovative management is the economic pulse of American health. It is the life blood for sustaining the strength of the economy, improving the quality of management, and securing the robustness of business. It is the fulcrum for the final fifth of the twentieth century. If America is to build a muscular national economy, benefiting all citizens and leading the world, the mechanism must include creative and innovative management.

Though words flow easy, precise definition is necessary. *Creativity* is the process by which novelty is generated, and *innovation* is the process by which novelty is transformed into practicality. Creativity forms something from nothing, and innovation shapes that something into products and services. To nurture and develop creative and innovative management is to engender America with the power to prosper.

Both collective policy and individual business are involved. If creative and innovative management can build industrial abundance in America, it will do so on two pillars: the macroeconomic environment and the microbusiness structure — macro and micro. But such flourishing will not happen by accident. It is a way of thinking new and hard. No one risks for little reward. Only within a proper environment will American management make the right moves and take the right risks. This environment has two elements: (1) an economic climate responsive to creativity and innovation and (2) a corporate culture conducive to such novel management.

BUILDING THE ECONOMIC ENVIRONMENT

Encourage Risk by Strengthening Reward

Proprietary ownership is a powerful human motivator; it is capitalism's great advantage over communism and we must pound it without pause. We should strengthen our patent laws now to include new forms of invention in the information and knowledge-based sciences. Government contracts should be structured to encourage recipients to reach and to risk, whether defense contractors, university science departments, or government laboratories. Both institutions and individuals must benefit from their toil. Federal research and development (R&D) funds, perhaps our nation's chief asset in building comprehensive national security, should embed economic as well as military forces, deriving maximally efficient value from each. Government contracts, for example, might be awarded to firms that generate original ideas or products or firms adept at commercializing defense-related technology, whether the firms be large or small. The current differentiation by size may be missing the mark.

Facilitate Information Transfer

Creativity and innovation are resources that increase with use: The more you use it, to quote Dr. George Kozmetsky, the more you have it. To enhance applications, we must publicize and promote. Although creativity and innovation are private processes, they can be fostered by information sharing and situation setting. Centers for Innovation and Invention should be established, funded by state and federal governments, and administered by colleges and universities. National data banks can enable active researchers and potential entreprenueurs to access ideas and information.

Focus Government Fiscal and Tax Policy

Many words are spoken in Washington; millions every year are written into record and law. None are heard more clearly, none are read more carefully, than those dealing with taxes. By tax law the federal government directs public policy. A clear message for developing creative and innovative management will be given only when tax policy is the medium. We should reward creative and innovative companies through lower taxes, rather than penalize their profits with higher taxes. Tax credits for incremental R&D is a first, albeit halting, step in the right direction. We might consider, say, tax credits for new patents, new products, and R&D expenditures above industry norms. Capital gains, as another example, might be dropped further, perhaps to zero, but only if, in my opinion, the holding period is increased. (The recent change to *reduce* the holding period to six months flies off in the wrong direction, encouraging financial manipulations not productive development.)

Understand the Creative Process

Public policy should support research and education in creative and innovative management. Studying the process should become a national goal, not a curiousity, a necessity. America's finest researchers should be funded and interdisciplinary work encouraged from organizational psychology and the decision sciences to artificial intelligence and the neurosciences. The arts, too, offer much and should not be neglected. In concert with research, we must stimulate creative and innovative management in our schools. Principles of creativity and innovation can be taught at every age, in parallel with enhanced math and science, from early education through high school and college. Schools of business should take the lead, instilling motivation to shift and change rather than drilling techniques to trend and continue. One danger of making business more rational, more analytical, and computer based is the subtle pressure to stifle the new and inhibit the fresh. Business executives must be prepared to make nonrational (not *ir*rational) decisions, gambling on instinct and perception. Though business should become more of a science, it must never cease being an art.

Promote Interaction among Sectors

Creative and innovative management is not sector specific. It occupies a unique place at the union of industry, government, and academe. Each sector must make its contribution, and critical mass can be generated nationally only when all focus their force on the interface. Intersector interaction is not just a current fad, it is the white-hot focus; and government policy should catalyze the reaction. The Department of Defense policy of rewarding companies with university ties higher scores for independent R&D funds is an excellent prototype. State government, too, must participate; they may, for example, offer matching incentives for state-based R&D, increasing operational leverage and financial appeal.

DEVELOPING THE CORPORATE CULTURE

Encourage Risk by Strengthening Reward

Most companies give mixed signals about risk. They praise new ventures with lofty words and reward failure with career wipeout. One such derailment incinerates the whole house of corporate cards. We must shift this risk-return tradeoff by decreasing the risk and increasing the reward. Incentives for originality and invention must be internalized and believed by the company underground. The organizational structure must support it; the informal networks must promote it; the grapevines must confirm it. Participating in new ventures — not just making them successful — must be the pinnacle of corporate achievement. "Have the Guts to Fail" is the motto of one innovative company. Creativity and innovation has expression, one should note, in all areas of corporate life, not just high technology and new products. Managers who look beyond the traditional, who see the unusual, who dare to be different — upon these does posterity rely.

Facilitate Creative Types

Egalitarianism, the belief that all are equal, is a fundamental American value. Although wholly appropriate in politics and society, it is counterproductive in economics and business. People differ in every respect, with the capacity for creativity at the top of the list. A company must respect its creative types. They are a breed apart, absorbed in their quest, dedicated to intensity, oblivious to others. Creatives are often difficult to control. They work strange hours in strange places. They don't want supervision and demand personal satisfaction for personal achievement. Proprietary participation, especially financial reward, is an essential motivation. How to find them? A word of caution. Creative and innovative people may not be the smartest or brightest; they may not be aggressive or assertive or even realize their own gift. The best firms will treasure them.

Focus Corporate Fiscal Policy

Companies that talk innovation and invest elsewhere dig credibility gaps. Promoting creativity is no mean task. A firm must evidence its commitment, putting cash on the line. Nothing energizes more than the movement of money. You can't talk creativity and fund tradition. The resource allocation process must encourage creativity and innovation; new procedures must skew dollars to more risky ventures. Most critical, results cannot be expected quickly. Corporate executives must see beyond the horizon, beyond the quarterly reports, beyond the street called Wall.

Understand the Creative Process

Creativity and innovation happens by itself, but not all the time. Since innovators are often not the brightest or most aggressive, the firm must find them, or, more accurately, help them find themselves. One cannot train people to be inventive, but one can develop educational programs to facilitate the process. Creativity appears with infinite variety. In a high-technology firm, for example, a person with a new method for inventory control may not think herself creative, yet the benefit to the company may exceed most scientific study. One good idea covers a lot of ground.

Promote Interaction among Divisions and Developments

Scientific advance depends on constant communication among diverse disciplines. Likewise for the best businesses. When problems are attacked by divergent approaches and disparate facts a wider range of solutions emerges. Task forces composed of different departments are not unusual in corporate life, but these are often established for coordinating current programs rather than creating new ones. Interdepartmental cooperation in companies, like interdisciplinary work in academics, is fraught with suspicion and worry about territoriality and dominance (the sociobiology of ant hills and wolf packs do not encourage creativity). A firm's new products division doesn't want manufacturing sticking its nose in; manufacturing says it's ridiculous to develop products that can't be made. Mechanisms must be found to break these barriers. The catalyst is often the person to whom the departments report; the boss must become actively and aggressively involved. If he or she "recommends" the interaction without personal participation, it will surely fail.

The opportunity is here, the time is now. What we have is nothing less than the restructuring and recrudescence of American industry. Economists and executives must work together in building both a macro/economic foundation and a micro/corporate structure. In the new realities approaching the year 2000, to achieve domestic vitality and world leadership, the American trick is creative and innovative management.

PART IV

STATE POLICIES AND INITIATIVE DEVELOPMENTS

10

The Role of State Government

The Honorable Mark White

I would like to give some insight into transportation problems that we share in the state. It took me 34 minutes to fly from Austin to Dallas and 42 minutes to drive from the airport to the University of Texas campus. Yet we have advantages in Texas that make it appropriate for us to get optimistic about the future.

Take the leadership of the University of Texas and Texas A&M University in technology venturing. Combining the two in computer science and electrical engineering makes an unbeatable combination. No one anywhere in the nation can match those resources, financially or intellectually. Years ago, we put together the Permanent University Fund. (Our predecessors probably weren't all that visionary at the time; there was just all this pretty land out in west Texas that was given to the universities. Fortunately, there was a lot of gas under that land and the accumulating resources have been dedicated to the intellectual advancement of our people. This is our springboard for future developments.)

Education is the central theme of technology venturing. Without the educational foundation, Texas would not have attracted the Microelectronics and Computer Technology Corporation (MCC) to Austin; they would have gone somewhere else. With our educational foundation, they came here.

Business growth and educational excellence are a dynamic combination. Yet, there is work to be done. Dallas is not focused as sharply on educational strength as it can and should be. Dallas has been a national business leader for years and has outstripped much of the nation in commercial development. It now needs to strengthen its educational and intellectual foundation. Conversely, the University of Texas has been a great benefit to Austin, helping to build our private sector and strengthen the business community. Educational institutions in Houston are now combining their efforts as well as those in Dallas and Fort Worth. We have strength throughout the University of Texas system statewide, and we must focus and chart our course very accurately to avoid duplication of efforts. We must make certain that we set a course of intellectual study and conduct that will be complementary not competitive. The old competition doesn't exist today between the University of Texas and A&M University. (We still maintain fierce competition on the athletic field of course.) There is no reason on earth for us to be scattering the scarce resources. Each university must fulfill specific short-term objectives and well-defined, long-term goals. We need to focus and make the right choices after setting the proper priorities; we must make sure that we do not waste our scarce intellectual resources for research and development.

This to me may be the single most critical factor that we face in Texas today. We have been blessed with many assets and for that reason we have sometimes been wasteful of those assets. But we must focus more on the realization that we cannot afford to duplicate our intellectual endeavors. We have to be well coordinated. We can set many priorities that will put national leverage on our resources.

What must state government do in order to maintain the momentum we are enjoying right now in Texas? I know no better example than that drive from Love Field. Thank goodness I was going this direction and not the other! If we have any lessons to be learned from the northeast and from Los Angeles it's the question of mobility. We, too, can become a joke. We, too, can fall into the same trap. We must learn the very simple lesson of maintaining mobility. We cannot be led down the primrose path as they were in Houston by a rather expensive but not particularly well-thought-out proposal for a transportation system that would not haul a whole lot of people but would soak up about $3 billion. Now that was the plan of some of the brightest people in the community. But when put to a vote, the people rejected it by 70 percent. Houston's electorate was asked to vote again for additional roadways and toll roads, and many of the more thoughtful political leaders thought that the proposal would go down just like the other one. No, the people *passed* it by 70 percent.

I think we sometimes forget that people have the inherent quality of knowing who is flimflamming them and who isn't. The problem in Houston was not that the trains were overloaded but that the roads were overloaded. You can't get anybody to get on a train. We are so smart, we think that we can arbitrarily design transportation systems that people will use. I'd hate to have to spend the rest of my life trying to figure out how to get people out of their cars and into a train. Nobody has been able to do it yet. (Because the energy crunch and because transportation was tied up, more and more people went to vans. But the first chance they get, they are right back into their cars.) What does transportation have to do with technology venturing? If employees can't get to work, you won't have any business.

Some of the key factors in MCC's choosing Austin were the mobility factor, the environment, and the quality of life. People who are well paid start making choices based on where they can enjoy their salary and family life.

Nothing is more important than education. We have all been talking about the University of Texas and A&M, but you know that there are many young people graduating from public schools in Texas that can't get into Texas A&M and the University of Texas. Why? Because they don't have the resources built into their primary and secondary educational life. We must demand more of our students. We must raise standards. Raise standards for admission to college. Raise standards for admission to the teaching profession. Raise standards for graduation from high school. I think our people will rise to the new standards and work hard to accomplish the goals we will set for them.

We will also renew the spirit of our public school system. Good cities and good schools go together. Invariable the answer to what happened to Detroit, New York, and Los Angeles is a declining public school system.

What do we need? More money for teachers? Certainly so. More math and science teachers? Absolutely! We are falling short in the number of math and science teachers. And if the trend continues, the only thing that we will be producing from our high schools will be young people who want to be lawyers — and we have enough lawyers already. Many students enrolled in masters and doctoral master programs are foreign, not American. And much of that technology we see abounding in Japan and Germany and other nations, we generated right here in our own educational system. But our people didn't benefit from it because our students just weren't in the program.

We also have problems with government in implementing technology venturing. Our antitrust laws are archaic: They do not fit the expanding role of the business community and what we are facing in world competition today with other countries. How can we beat that competition? The MCC consortium, a multitude of companies joined together in a collaborative research effort, is a unique challenge to the Sherman antitrust law, and something that needs to be replicated. It is something our antitrust laws should recognize as being a valid combination of competitive talent. How you go about sharing that technology, and how you keep the door open so that it doesn't become a closed shop or a monopoly, is a balance that can be handled. It is important to open technological development to all comers though there is some preferential treatment given to the investors. The economic realities of the world marketplace are changing, society is changing, and so must government.

As Americans we must rise above state vs. state competitions. We can no longer compete against each other — Texas versus California or Texas versus North Carolina. We can't afford the luxury. We are in world competition. If you don't think so, then go out in the parking lot and see how many cars came from this country versus those from other countries. And the reason for it? Check out the way we organize management and labor. We need a new relationship between management and labor, a new recognition of mutual interests. We must make innovative changes in our means of operations to keep jobs in America, make workers more productive, and make certain that our products are competitive in the marketplace.

Americans are winners, and we are absolute winners. We can outperform, outproduce, and outinnovate any other people on the face of this earth. We have created and epitomized the entrepreneurial spirit. We have shown it can be done in the past, and it will be done in the future. We have created and epitomized the entrepreneurial spirit. Technology venturing is part of the process, and what this volume does is support the state and benefit the country.

11

Social Counterpoint

The Honorable Henry Cisneros

What are the implications of technology for the challenges of American society? Technology as an instrument of American progress is a means toward the end of American ideas. Not ideas that are strictly technology ideas, but ideas that are decisively nontechnological, such as equity consideration, justice, quality of life, upward mobility for all our citizens, and the maintenance of a society that is open, mobile, and lacking in the rigid class distinctions that characterize other societies in the world.

We must consider technology within the fundamental ideals of American society. I will play the role of devil's advocate juxtaposed against some chapters in this volume.

Why is it important that people with technology leadership in business, government, and academia think about social issues when evaluating technological progress? There are several reasons. One has to do with the array of political forces in the society, and a second one has to do with institutional gridlocks or constraints. The failure to understand political forces or the nature of institutional blockages will trigger some traps.

What are some of the political forces that might be contrapositioned against technology issues and imperatives? Certainly there are environmentalists with automatic skepticism about projects of scale. There is labor, with their justifiable and legitimate employment concerns. There is the political movement concerned with simplified technologies and small-scale activity. There are interests of certain industries, particular basic industries that want to protect a *status quo.* There are minority groups, ethnic minorities in the country that are not sure how they relate to the technological process.

All of these have a certain political following and constituency, and consequently, a certain political force and pressure. This is important because given the nature of our society and the institutional processes, those constituencies can mobilize constraints, establish blockages, and enforce stalemates that pose gridlock. They can use, for example, the congressional funding procedures and schedules and the regulatory process to create political deadlocks.

This is the shape of the world, not necessarily the dominant shape of the world or the decisive shape of the world but an element of the environment in which technology venturing must occur.

It is important to understand some of the things that concern these constituencies and to anticipate them. Perhaps the technology community should address these issues.

What are some of the critical issues before the society that involve technology? The first is jobs. Perhaps the central issue before the American people for the rest of the 1980s will be jobs. I do not have to detail the transformation of the U.S. economy: an economy in the early 1950s in which 60 percent of the people worked in industrial jobs and 20 percent of the American people worked in handling information, to an economy in the early 1980s in which about 20 percent of the American people are employed in industry and about 60 percent in handling information. Displacement has occurred. In the depths of the recession in 1982, General Motors had laid off something like 160,000 people. During the same time period, the company ordered 14,000 industrial robots to be put in position over ten years; 65,000 people probably would not be asked back to their jobs. We will not see an employment rate below 7 percent for the rest of this decade. Full employment is still a far-off goal.

Now 7 percent is possibly justifiable on the grounds that the makeup of the American labor force is different today. There are different groups of people in it who we never counted before in a full employment economy that are today counted. There are many other changes in demographics, and as a result it may be that our society can tolerate 7 or 8 percent unemployment for an extended period of time. But it is a relatively new dynamic to sustain that level of unemployment and it will affect our social fabric.

Also critical is education. We need some dramatic changes in public attitudes with respect to education and its role in technological change. A society that has not fully prepared its students to understand the three R's, reading, writing, and arithmetic, will have great difficulty in transforming itself to teach the three C's, the ability to calculate, to compute, and to communicate in the language of technology.

Education is a public issue. But the reason why it becomes a particularly sensitive public issue is for reasons of equity. There are divisions in American society today. Those divisions too often take the lines, unfortunately, of racial and ethnic divisions. The statistics on unemployment, the statistics on income and poverty for example, break along those lines all too easily. And yet I worry that ten years from now, these divisions in American society may be even deeper. Today we can use tools such as affirmative action programs, and compensatory education. But what ten years from now when the lines aren't as easily crossed? When chasms separate those who are in the productive sector of the society from whose who are not. Or the fundamental difference between technological literacy and illiteracy. Technological incompetency will not be able to be cured by a quick six-month compensatory education program that can cross a barrier rooted in the failure in the sixth grade to master science, math, and analytical thinking.

There are many sectors of society affected by technology expansion. Will there or will there not be enough jobs? What about educational preparation for displaced workers? And what about the issues of equity? What about wage scales in industries in decline, and when different groups of people are forced to accept lower paychecks. Serious questions for society. Not insoluble or inevitable. But leaders in technology need to be addressing them.

Unanticipated problems can only generate unnecessary blockages to technological progress. Consciousness of the jobs issue, rather than dismissal of it as something that will take care of itself, is a critical challenge before American leadership.

A second issue with a technology dimension relates to the institutions of society, specifically the institutions of government necessary to handle a population of 250 million people, to move the American polity. In a period of economic transformation, we need to begin institutional transformation. Texas is a good example.

Texas in 1983 suffered a triple whammy to its economy base. Some would say that the shot that afflicted Texas was not triple but really had four dimensions. The first was the lag effects of the national recession as it reached Texas in early 1983. The recession had its trough nationally in October and November 1982. Though it took a little while before the recession was felt fully, since the Texas economy today (with a large manufacturing component) is more like the national economy than ever before, the recession did hit and hit hard.

A second dimension of that triple whammy was the turndown in oil prices. One need only study the oil patch in Texas, Midland, Odessa, Beaumont, Port

Arthur, Houston, and other areas indirectly connected with oil, either in production activities or in oil field equipment manufacturing, to understand the magnitude of the problem. Houston, which had weathered every previous national recession, sustained unemployment approaching 10 percent in the summer of 1983 when oil prices were at their lowest point. Houston suffered in 1984 an unemployment rate of about 8 percent, which was twice as high as the Austin and the Dallas-Fort Worth areas.

The third dimension of that shot on the Texas economy was the peso devaluation in the borderlands. It is difficult to describe in words the human drama of 27 and 28 and 30 percent unemployment. But that is what hit Laredo, Eagle Pass, and Del Rio and parts of the valley this last year. People literally had to turn in their car keys and their house keys because they simply weren't able to make the payments after unemployment compensation benefits ran out. That was the story of the borderlands in 1983, and unfortunately it's a story that continued because of a fourth shot on the Texas economy: the weather.

The bolt from the skies took the form of a serious summer drought on the high plains that affected livestock and agriculture, and a subsequent freeze in the valley that has unemployed some 30,000 people as a result of a reduction in direct agricultural work and in processing. One processor of orange juice, for example, had to let 300 people go.

So 1983 was a year in which Texas leadership had to come to a basic recognition that the "invulnerability" of the Texas economy, the so-called recession-proof Texas economy, was indeed a myth.

Luckily, 1983 was also a year of institutional transformation. The Microelectronics and Computer Technology Corporation (MCC) came to Austin. It was a great victory for Texas as 57 other cities across the country cried for the research consortium that is described in Chapter 5. But perhaps the most significant thing about MCC for Texas is not going to be the number of jobs directly associated with it or even the jobs spun off or other companies attracted. It is likely to be the institutional change within the state.

A governor for example who employed his office as no Texas governor before him had done. He utilized the office of governor as a command center for the direct competition with other states. He brought the chancellors of the major university systems together and said in effect that this state is not wealthy enough for its major institutions to pull in their own directions but rather they need to pull together and integrate their strengths in this endeavor. And that was done. The governor organized an office of economic development within his own office and reconfigured the Texas industrial commission to become the Texas economic development commission; he employed a whole new array of tools and assigned new missions to that commission. In 1983, the governor made the link between education at all levels and economic development as it has not been made before: primary, secondary, and higher education, each with important roles in the quest for jobs in Texas.

What we witnessed then was *institutional change* that begins to address the kind of changes that will be necessary for Texas to compete in the technological era. The kinds of institutional changes that had preceded Texas in North Carolina, Massachusetts, and California. There is no doubt that we are going to see institutional change in the private sector, for example in the capital institutions, banking, and finance. But I think the real challenge is going to be government; the bureaucracy of state and federal agencies must transform itself in this period of economic transition.

The third issue in which technology has a role is the continued process of decentralization of American society. In his best-selling book, John Nesbit talked about ten megatrends sweeping America. Larger than any personality, larger than any piece of legislation, larger than any political moment, they are the soul and psyche of the American people. One megatrend was the decentralization and specialization of American life. Not just governmental decentralization, but in the private sector. Take for example, the specialization of the publishing industry: What once was centered on a few major news magazines, Life, Look, and Saturday Evening Post, is today characterized by 100 choices: women's issues, money issues, management, psychology, cameras, airplanes — whatever your specific interest. Such decentralization is invading in broadcasting. An industry that once was three major networks, today can be characterized by 30 or 40 options, even in small cities. Because of the cable, because of the satellite dish, people can choose all music, all weather, all art, all religion, all talk, whatever it is that interests them at a given moment. The significance is that people are hearing different things, seeing different things, and, therefore, thinking different things. What works in Brownsville, Texas, won't work in Fresno, California, and neither is likely to work in Boston, Massachusetts, or in Minneapolis, Minnesota.

It really is an era in which a good deal more latitude, fervor, innovation, and imagination are occurring in regional settings. The action, if you will, is local. In regional universities, in small business, in state and local governments, and school districts, we see innovation in forms of governance. The corollary is that cities and communities that plan will do better than those that do not. Cities and local institutions that acquire a strategic sense about themselves will do better than those that do not understand this concept. Cities, throughout history, have been able to influence their destinies by making conscious decisions shaping their relationship to the larger trends and forces in society.

One could cite for example, the development of New York and its port. In our early history, New York was not the prime port for the United States, Boston was. But Boston was characterized by a relatively nonfertile hinterland, the rocky land and cold soil of New England. New York had no great access to richer lands until the decision to build the Erie Canal. The result was the opening up of produce-rich areas to the port and New York quickly followed as the dominant port.

One could cite a more recent example in Texas around the turn of the century when Houston built its ship channel. The idea that the Gulf of Mexico and Houston could be connected by ship channel made possible the development of the petrochemical complex in the Houston area and the city's eventual development as the world capital of oil technology.

A still more recent example was the effort in the 1960s, lead by Mayor Eric Johnson of Dallas, to set long-range goals for Dallas. The result of this strategic sense was the realization that without a replacement for Love Field, without a facility that could be "a port to the ocean of the air" as it was described, Dallas would have to compete with Tulsa and Oklahoma City and with other southwestern cities for a dominant role in the southwest. The decision to build DFW Airport, in my judgment, guaranteed Dallas' role as the capital of the southwestern states in distribution, finance, and wholesaling and thus assured the city's continued economic position among the strongest in the country.

The examples do not apply only to large cities; small cities can make decisions that determine their destinies. Lowell, Massachusetts, a case in point, lost textiles and other manufacturing in the 1950s and 1960s, and in the 1970s decided to concentrate on a major downtown restoration in conjunction with Wang Laboratories. The result is that Lowell is today one of the thriving small cities of the industrial northeast. Fort Wayne, Indiana, lost major tractor companies several years ago. Two years later the city was named the most livable small city in America and an all-American city by President Reagan for its goal-oriented response. The point that I am making is simply that in an era of local initiative, cities that have conscious ideas about where they want to go will be more successful.

In my own community of San Antonio, we've started a process that we call Target 90. It is an opportunity for 400 leading citizens to think through what we want to be in 1990. We are considering some 170 objectives that were preselected, but not limiting ourselves to those. We are looking ahead less than 60 months from this moment, not in Buck Rogers terms, not in pie-in-the-sky terms, but in specifics: what expressway lanes need to be widened to avoid the congestion, what our SAT scores should be, what new curricula need to be put into the local higher education institutions and into the high school vocational programs; what is the state of our arts institutions, and what levels of funding have to be achieved? These are the kinds of things that the Target 90 effort speaks to.

We have identified five areas of technological jobs that we want to target: industrial processes equipment, electronics and telecommunications, aerospace and defense, agribusiness technologies, and the biosciences. We have learned that there is an almost mathematical precision about the business of economic development. If the elements are not in place then they are simply not in place. The equation does not equalize. If the energy is not available, if the water is not available, if the political climate is not correct, if the incentives are not in place, then the equation never equalizes.

But what is interesting is that today the terms of the equation are new. What once used to be the required infrastructure for the growth of cities and towns across this country — easy access to seams of coal, water born transportation, the availability of port facilities — has given way to the new required infrastructure of the technology era. What are those new requirements? The availability of venture capital; the readiness of the local community to invest in education at all levels; the creation and research park environments in which companies can locate; the access to a critical mass of research activity; a quality of life in which research personnel can make an attractive life; the support systems for export development; the creation of incentives for foreign investment; far-sighted state policies on technology; and finally the projection of an image that says we can create the environment in which technology jobs can flourish. I submit that relations between technology companies and the state and local governments are going to be very important as we move through the rest of this decade.

The final issue in which technology has a role, but as yet an undefined one, is the whole set of issues related to the challenges of our American foreign policy and the appropriate relationship with those parts of the world that have never been part of our foreign policy interests. I served on President Reagan's bipartisan commission on Central America.

The opening chapter of our formal report says roughly the following: For the better part of 200 years, the American people have given their best face to Europe and the Atlantic Alliance for reasons of culture and history. Europe has been the central theme of our foreign policy. Indeed we have fought two world wars there to defend the countries. We created a defense system, NATO, and rebuilt Europe after World War II with the Marshall Plan. For the last 50 years America has focused on the Pacific, and the names that have made the headlines include those of Japan, Korea, the Philippines, China, Vietnam, and so on.

But as we approach the last decade of this century, it is increasingly clear that we can never again be apathetic to problems of the countries to the south of us, Mexico, Central America, Latin America. Despite the geographic proximity, we have been far apart.

One cannot discuss Latin America without discussing its incredible debt. Commodity prices such as coffee and bananas dropped in the world recession at the very same time when the need for social spending increased bank borrowing so that in some instances 40 percent of national budgets are going to pay the interest on loans to American banks. A grinding poverty is the legacy of 300 years of oligarchic rule, legal institutions that don't work, internal strife, illiteracy, poor health, rebellions, and revolts in the region.

I submit that we will not be able to find answers in technology. What is needed is not just exports, not just the standard processes of technology transfer. We need people to build bridges to the disciplines and the institutions able to solve those most complex problems: problems that are not solvable by military solutions but that have a security dimension; problems that are not solvable just through aid for education and social services but for which aid is critical and necessary.

I believe technology is the driving force in American society. It should be and will be. But I do expect that there will be constituencies and constraints and blockages because technology will not be able to address the kinds of questions I have cited.

The jobs issue, for example, has not been addressed in forthright ways, particularly by leaders of technology. The transformation of American institutions, particularly governmental ones, must occur. Many things will change but at least one thing must stay the same, the basic American characteristic — optimism. If you scan American history, the notion of optimism is constant throughout all periods — the hope and expectation of a better future — the dynamic that makes two plus two equal five because of the lifting effect of optimism.

I close with a vision from perhaps the darkest moment in American history. It was not the 1980–82 recession, it was not World War II in the bleak beginning months after Pearl Harbor. The darkest moment was not even the Great Depression when 30 percent of the American people were out of work, and soup and bread lines were ubiquitous. The darkest moment in American history was the Civil War, when the country threatened to tear itself apart. It was not clear that the Union would survive. And yet in that period, the Congress of the United States met in Washington and passed a remarkable piece of legislation that affects us all. That piece of legislation was the Morrell Act of 1862, which created the land-grant college system. And whether you attended one or not, you've eaten the beef that has been prepared at these institutions and benefited by the machines designed by the engineers trained there. The truth of the matter is that in the darkest moment of the country's history, when it wasn't clear that the war could be won or the country would survive, the Congress made a statement of optimism dealing with education.

Education, by definition, is an act of optimism. It is investment in the future — investment of time, money, skills, and talent in a future that presumably will be better. The Congress said in effect, "Go out and train the engineers, because when this war is over, we are going to have to build dams, and roads, and bridges, and towns. And train the animal scientists and the agricultural economists, because when this war is over, we are going to build a farm-to-market system, and we are going to have to feed people and teach teachers and prepare the nation for rebuilding. We are going to build a nation."

That spirit of optimism has always been part of American history, from that darkest moment through the present. In 1969, barely a decade after the shock of Sputnik, an American walked on the moon. And today, confronted with international competition and other problems, the American response once again will be one characterized by a sense of optimism. It is part of the American nature and I thank God for it.

12
State Legislative Initiatives

Joseph Kigen

What are the roles of the states in this great national dynamic of high-technology venturing, and specifically what initiatives are unique to them in constructing public-private partnerships?

Clearly these public-private partnerships are the institutional plasma that will breed and nurture an achievement level for America reestablishing our preeminence in world commerce and removing us from the morass of "dinosaur industry."

World commerce is a complex issue — several continents, many nations, countless provincial or state jurisdictions of both lawmaking and commerce, and they are all interdependent. None is a point of isolation, particularly when the stakes are so high in the race for global technology superiority.

Given the awesome complexity of high technology, given the inordinate costs of program development, given the sophistication of applying the awarded gains in innovation, it seems like a contest in which only federal agencies of government are capable of competing.

I suggest that this thesis is wrong and that intergovernmental process is the order of the day. Just as the German Bundestag is dependent on the state of lower Saxony to develop a national perspective on critical political issues, just as the People's Republic of China is giving new economic and trade-stimulation independence to the province of Sichuan, just as the federal government of Brazil so attentively respects the state of São Paolo whose gross national product (GNP) matches that of the nation of Argentina, I think the U.S. Congress and this or any administration will welcome — and need — the support of America's 50 states, as individual lawmaking bodies and as a collective voice of our nation in high-technology policy development.

Let's explore several examples of public-private partnership in action at its very best.

It surprises many to learn just how active the states have been in the high-technology arena in recent years. I must pay tribute to an excellent summary of this activity that was prepared by the National Governors Association (NGA) as the Final Report of the NGA Task Force on Technological innovation, which was chaired by Governor Jim Hunt of North Carolina.

The NGA report comments on state high-technology activity in five broad program areas: (1) policy development, (2) education, training, and employment, (3) linking university/industry research, (4) technical and management support, and (5) economic support and incentives. I will stress particular program areas that most closely relate to the goals of this volume.

Although Texas typically has a leadership posture in many areas of commerce, it certainly can respect the accomplishments of its peers in terms of what they have done in high technology, championed at both the executive and legislative levels. The dynamics of this great American "high-technology sweepstakes" are fascinating.

In the policy development area, there has been a great deal of activity in the state capitals. In the last three years, governors in 27 states have appointed special advisory bodies with the specific task of analyzing and reporting on state capabilities and needs and recommending programmatic solutions to promote technological research and development and industrial expansion. Over two-thirds of these advisory bodies have been designated as "high-technology" or "science and technology" task forces.

Some of these are not located in states that are commonly viewed as being in the Olympic games of high technology. For instance, the state of Iowa recently established a permanent High Technology Council and awarded it an initial research budget of $2 million.

In states with a more visible high-technology base, the commitment is more dramatic. In Washington, a gubernatorial committee recommended a $65 million program expanding engineering education, augmenting technical and teacher training, and creating a high-technology center. One immediate implementation of these recommendations was enactment of two laws providing $13.5 million for customized job training and high-technology education and research.

The state of New York traces its Science and Technology Foundation back to 1963. Having a broad charter to promote statewide science and technology education and basic and applied research and development, the foundation serves as the conduit between the private sector and the academic community to promote joint research and development (R&D) ventures. It now has an annual budget of over $4 million.

In a number of states, the agencies responsible for economic planning and development are expanding their traditional focus and outreach to include specifically technology and industrial innovation.

The state of Pennsylvania, through what is known as the Ben Franklin Partnership Fund, has established four advanced technology centers, each operated by a consortium of business, labor, financial institutions, academia, and local development districts. Each of these centers is eligible for 50 percent state support, with matching money from other sources. Pennsylvania's commitment to the Ben Franklin Fund for this year is $10 million.

Massachusetts has created a state Technology Park Corporation, which is authorized to establish a statewide network of facilities dedicated to specific areas of science and technology. The first facility proposed under this authority is a $40 million Massachusetts Microelectronic Center. Half of the funding for the center will be state source, with the balance provided by a consortium of private companies and academia. Governor Dukakis has also committed both support and state property for development of a Biotechnological Research Park.

Among the several states that have set up nonprofit, quasi-public institutions to oversee that high-technology development, one of the more prominent is the Tennessee Technology Corridor Foundation. This group would oversee development of an actual eight-mile corridor from the Oak Ridge National Laboratory to the University of Tennessee campus and TVA (Tennessee Valley Authority) headquarters in Knoxville. Established with a $1.3 million commitment, the corridor is planned for concentrated development of high-technology business and allied industries.

The NGA report also points out the existence of private advocacy groups comprised of senior executives from advanced technology and related companies. These groups have established themselves in some cases as effective lobbying arms of the high-technology community. Witness in Massachusetts the success of its advocacy group in championing a reduction in the state capital gains tax and in lobbying for the ballot issue that substantially reduced state property tax.

The Washington State Council for Technology advancement recently secured passage of legislation modifying sales tax provisions for R&D prototype development and eliminating taxes on corporate donations of computer equipment to educational institutions.

In the area of education, a number of states have begun to address the problem posed by declines in elementary and secondary education. State policy and planning advisory groups are being set up, closer public-private partnerships are being established, and specific measures are being taken to meet the broadly determined needs of schools.

One example in this area would be the Minnesota Alliance for Science. Composed of business, labor, and academic leaders, it seeks to strengthen the quality of science and mathematics instruction in elementary and secondary levels. The alliance's formation followed public awareness and shock over the revelation that, by 1991, Minnesota's nearly 2,000 technology-intensive businesses would have to import 80 percent of their engineers from other states unless Minnesota's own output of engineering graduates was increased.

There are also efforts in some states, such as North Carolina, Louisiana, and Oklahoma, to establish programs for accelerated math and science instruction for gifted students.

Recognizing that effective worker training is a key for sustaining strong economic growth, a number of states are instituting programs to ensure an adequate supply of skilled and highly trained workers to meet the demands of a technology-based economy and to provide retraining for those whose skills have become technologically obsolete, the unemployable veterans of the rustbelts of America.

Thirty-three states have programs of "customized job training" to meet the needs of specific industries. One showpiece example of this is Missouri's new General Motors Training Program. This is a cooperative effort of the state, GM, and the United Auto Workers to train workers in robotics and related technology skill areas for the opening of a new GM assembly plant. It is ultimately hoped that some 2,000 hourly workers will be trained as technicians to operate the plant.

In the area of higher education, several states are earmarking it as a high-priority resource in achieving high-technology parity. At Arizona State University, for instance, a new Center for Excellence in Engineering is under development to support education and research in numerous high-technology disciplines. The center, funded by a combination-source total of $33 million, is scheduled for completion this year.

The state of Washington is funding a $20.6 million cooperative program, the focal point of which is a new High Technology Center at the university with special emphasis on bioengineering and microelectronics. The program also includes a statewide telecommunications system to provide graduate education programs in engineering and management.

On the subject of linking university and industry research, we find an area of vigorous participation. In 1983 alone, industry spent some $300 million on scientific and technical R&D in academic institutions and libraries. Although this figure is small compared to the $7 billion government funding of academic research, it does represent a fourfold growth over the past decade. There is also an increasingly evident role of state government becoming involved with industry in joint funding of new research facilities.

In 1980, 74 percent of all business support for university research came from just five major industries: oil and gas; chemicals; electrical machinery; food, beverage, and tobacco; and pharmaceuticals. One phenomenon contributing to this has been the increasing interest of industry in targeting university research, both locally to areas of commercial interest to that industry and broadly to wide vistas of technology.

Another concept of great resource value to high-technology growth are research parks. With their well-integrated clusters of industry/university linkage, research parks either exist or are in some stage of planning and development in 18 states.

The most prominent and largest of these is Research Triangle Park in North Carolina. The product of nearly 30 years' effort, Research Triangle comprises 5,500 acres and contains more than 36 separate research facilities located within the triangular perimeter of North Carolina State, Duke, and the University of North Carolina. This is a showpiece display of the efforts and persistence of the private sector, citizen and state support, and balanced interplay with a solid academic structure in the state. Another program area of the NGA study showing widespread state activity is technical and management support. Focused heavily toward small businesses within the states, these programs provide effective avenues for firms to deal with the possible trauma of commercialization of technologies into the public sector.

"Innovation Centers" have become increasingly popular as tools of both states and academic centers to provide management and technical assistance to small technology-based businesses. One of the more interesting of these centers is the Indiana Institute for New Business Ventures, established just last year. The institute will focus on promoting economic development and new employment in Indiana through the formation and support of small, technologically oriented new business enterprises. It will serve as a broker linking new business ventures with the technical, managerial, financial, and educational resource capabilities in the state.

Research incubators are another promising tool being employed by the states to provide a suitable environment for "hatching" new ventures spin-off applications of high technology. They are an essential part of the comprehensive technology programs of Georgia, New York, and Pennsylvania. Incubator facilities have also been created in Illinois, North Carolina, and Oregon.

To the innovation programs and research incubators, a number of states have also established training and support programs for entrepreneurial development, feeling that this is one of the most critical areas in promoting technology commercialization.

The last program area cited in the NGA report is that of economic support and incentives. A number of states are now developing special initiatives such as tax incentives, enterprise zones, research and industrial parks, and direct financial assistance in the form of low-interest bank loans and loan guarantees to fulfill the start-up capital needs of new-technology firms.

One area of concentrated interest is state support and implementation of private venture capital sources for small high-technology firms. Given the high risk and generally long-term payback time of typical venture capital investments, firms with less than $500,000 "start-up needs" are not attractive investment targets for the private market.

A fine example of state involvement in this area is the Massachusetts Technology Development Corporation (MTDC). Operating both as a public agency and nonprofit corporation since 1979, the MTDC has invested $3.4 million of its own funds into 20 individual firms. More dramatically, it has used these investments to leverage more than $22 million in private sector money to these same companies.

Due to the disproportionate share of venture capital availability to only a few states, several state legislatures have recently passed laws to aggressively attract a venture capital base within their borders. Nevada, for one, has a new law that overturns previous restrictions on venture capital companies. Montana has a new "Capital Companies Act" also aimed at stimulating in-state capital growth. This statute provides investment tax credits to individuals of up to 25 percent of their investment in private venture capital companies that focus on high-technology borrowers. Individual taxpayers could earn up to $25,000 for investing in those companies.

A creative approach was taken by the Michigan legislature in 1982, permitting the use of Michigan State Retirement Funds to be used as a venture capital base for equity investment in Michigan High Technology firms. The estimated value of these venture funds is $350 million.

Now, what has Texas done and what might it contemplate doing both in the near and long term to stimulate the growth of high-technology businesses?

Texas has been active in technology stimulation, but not as much as it should have been. It can speak of achievement in high technology, but not the depth of achievement that might be expected from a state that is traditionally in the forefront of business promotion activity.

In the last two years most of the technology stimulation activity in Texas was done without the participation of Texas government. It has been solely in the domain of the private sector. One of the accomplishments was the establishment of the Houston Area Research Center (HARC), which was established by major universities in Texas. HARC will oversee grant and contract research for private industry, government, and other groups.

Another Texas effort was the establishment of INVENT, the Institute for Ventures and New Technology at Texas A&M University. This was initially funded with a $1 million appropriation. INVENT has a specific focus on creating new business in the state, particularly those that use inventions and creativity as a feedstock for growth. It assists inventive entrepreneurs, particularly small firms, in developing business and marketing plans, and it helps them complete research prototype testing and design feasibility. For its efforts, INVENT charges a royalty fee or takes an equity position in the firm.

Probably the most gainful and prestigious new resource for the high-technology community of Texas is the presence of Microelectronics and Computer Technology Corporation (MCC) in Austin. MCC, under the direction of Bobby Ray Inman, will hopefully be for Texas and for the national technology base of the country the bedrock of a new wave of commitment and intensity of research and development that will promote American industry to a position of preeminence in the global interplay of high technology. But MCC, however self-sustaining it may be, needs the support and cooperation of Texas government if it is to reach its potential.

MCC's presence and the traditional dynamics of Texas entrepreneurship, coupled with the superb university structure in the state, are the ingredients for vigorous high-technology development. But the needed icing on the cake is the entry of Texas government into the public-private partnership.

There should be a careful examination and analysis of where deficiencies exist in Texas law or regulatory authority that inhibit technology growth or industrial expansion or job creation. And, Texas is blessed with some unique tools for the analysis and recommendation: the excellent system of the University of Texas and the public colleges and universities in the state; the presence of the IC² Institute, the entrepreneurial spirit of Texas industry; the presence of MCC as the now-established embryo of the public-private partnership concept; and the interest of Texas governor, Mark White, in contributing to the prosperity and vitality of the state.

Concurrently, Texas should use its own creativity or borrow from the success of other states in establishing positive incentives to fertilize the dynamics of business growth. If this means alteration of the revenue base, if it means creation of a loan fund or loan guarantee program, if it means the dedication of pension funds for venture capital purposes, if it means the establishment of technology incubators or innovation programs, then these things, and much more, should be done.

And if short-term political risk stands in the way of these technology-enhancing programs, then it should be overcome by the realization of the ultimate long-term benefit that will accrue from enlargement of the state's revenue base and the competitive enhancement of its businesses.

The gains seem obvious for the state of Texas, its private sector, and its working people. In a number of other states, they have been incubated and nurtured and now are being proven. And they will work for Texas, particularly by intensifying the now-established public-private partnership to match the cadence and drumbeat of the high-technology march through the 1980s. This is a great planet, this earth is ours, and we all — the states, the federal government, and the private sector — have a golden opportunity to leave it a better planet, through the legacy of partnerships in high technology.

PART V

LARGE-SCALE PROGRAMS: PROBLEMS AND OPPORTUNITIES

13

Toward A U.S. Technology Agenda: Insights from Third-World Macroprojects

Kathleen J. Murphy

The decade of the 1970s will probably be regarded as the watershed decade for the developing economies for many years to come. Beginning with the oil embargo of 1973 and continuing through the second oil crisis of 1979, enormous volumes of capital, technology, management expertise, and capital goods flowed to Third-World economies from western countries. Feasibility, design, engineering, construction management, and complex process technologies and skills were made available to government ministries and parastatal corporations on a competitive bidding basis. U.S., European, and Japanese companies transferred skills in such a thorough and professional manner that numerous local companies and career professionals have sprung up throughout the developing world.

In essence, large-scale development projects provide the front-end basics upon which the rest of the developing economy can take root and flourish. The sample of macroprojects (over $100 million in investments) include hydrocarbon processing projects, 23 percent (that is, oil refineries, gas processing, LNG (liquid natural gas), LPG (liquid petroleum gas), coal mines, synfuels, oil/gas pipelines); metal mining, refining and smelting, 18 percent; large-scale manufacturing, 6 percent (including pulp and paper, cement, auto manufacturing, textiles, etc.). The remaining half of the sample is composed of various infrastructural development projects, such as power, transportation systems, urban development, agriculture and irrigation, and industrial cities (6 percent). During the 1970s alone, these projects totaled 1,615 in number, and $1,010 billion in planned investment.[1]

Kathleen J. Murphy, *Macroproject Development in the Third World* (Boulder, Colorado: Westview Press, 1983).

U.S. companies were well positioned to provide the needed technologies and engineering and construction services to these world-scale undertakings, many of which were the largest known to date for the particular project industry sector. In fact, the notable successes of U.S. firms provides an excellent example of the dimensions of global demand for U.S. technology and management expertise. The unique contributions of U.S. companies, vis-à-vis competitors of other nations, confirms that technological leadership — and preeminence — are critical factors in effectively competing in the global economic arena.

As exciting as it was to collaborate in the design and implementation of Third-World development projects and programs, the rush to develop seems to have been more a temporary "heaviness on the gas pedal" in an overall slow journey, rather than a permanent change in the speed limit. More worrisome is the fact that the very market conditions and customer needs that underpinned U.S. successes in the Third World seem to be changing.

Although past achievements were realized with minimum, if any, national-level promotion and support, it is time to consider whether the formation of a national technology development agenda might not be appropriate. Governments of European nations have been quite effective in mobilizing local companies to join together in exporting their technologies and services. The list of U.S. companies who are active on major projects in the international arena is actually quite small. National level initiatives could broaden the base of corporate participation.

As a background for such a discussion, this paper draws on a survey of macroprojects in the Third World to highlight:

1. The proven strengths of U.S. engineering/construction firms in the international arena.
2. Changes underway in the macroproject marketplace.
3. The technology venturing option as an effective response to changing competitive conditions.

STRENGTHS OF U.S. ENGINEERING/ CONSTRUCTION FIRMS

The macroproject sample includes both industrial and infrastructural projects, a distinction that tells more about who the project sponsors might be or how the project might be financed than about what kinds of engineering or construction services will be required. A more fitting dividing line might be between those projects that require that technologies be tailored to local geological and other factors and those projects that require processes that are installable on a turnkey or plant-export basis, with minimal tailoring to specific local conditions. U.S. companies have been able to distinguish themselves in both areas.

U.S. design and engineering firms have been able to distinguish themselves by providing specialized engineering capabilities (such as Metcalf & Eddy and Engineer Science in the water supply/sewer system areas) or by providing diversified engineering capability (such as TAMS, airports, railroads, and hydroelectric). U.S. providers of process technology offer leading hydrocarbon processing and metal processing technologies, often on a fixed-fee or turnkey basis. These firms, though providing excellent services, do encounter head-to-head competition from European and Japanese and other companies in the macroproject arena. A really unique feature of U.S. engineering/construction capability, however, is the ability to offer project management and construction management services, particularly for extremely large-scale projects, along with their design or process. The extent to which U.S. companies are awarded these types of contracts distinguishes them from all other groups of competitors. It is unclear whether this is an outgrowth of the large scale of typical U.S. projects; or whether this is a direct outgrowth of the project management reputation and expertise developed by NASA during the space program of the 1960s.

The Engineering/Design Firms

A review of the U.S. companies participating on more than one macroproject in the Third World during the 1970–83 period shows that the engineering and design field is in fact quite concentrated and specialized. The list of leading firms is actually quite small, yet the investment value of projects that they have been invited to engineer or design represents billions of dollars and an enormous sum of man-years to implement. In terms of services delivered, each firm offers a variety of services including feasibility studies, engineering consulting, and design. It should be noted that many of these leading firms have participated in design consortia; however, U.S. firms team up with foreign companies more often than with fellow nationals. Contract awards to manage or supervise construction are also common (Table 13.1).

TABLE 13.1
LEADING ENGINEERING/DESIGN FIRMS
(Macroprojects 1970–83 only)

Specialization	Name of Firm	Number	$MM	F	CE	D	Dc	D/CM	DC
Road, railroads,	Parsons, Brinkerhoff	4	3,952	2		1	1		
metros	TAMS	3	2,269			2	1		
Airports	Airway Engineering	3	607	3					
	Bechtel	2	3,250	1				1	
	R.M. Parsons	2	3,151				1	1	
	TAMS	7	2,585	5		1	1		
Ports/harbors	Soros	2	310	1		1			
Water supply/	Camp Dresser McKee	3	1,300	1			1	1	
sewer system	Engineer Science	6	1,838	1		1	1	3	
	Metcalf & Eddy	4	2,389	1	1	2			
Urban	The Architects Collab	3	700	1		1	1		
development	CSR Design	2	517	2					
	DMJM	4	1,825	1	1	4			
Hydroelectric	Harza Engineering	8	22,473	2	2	1	3		
	International Engineering	3	12,200				3		
	Chas T. Main	8	3,720	3	1	3		1	
	TAMS	4	3,144			1	3		
Thermal	Gibbs & Hill	3	829			3			
Nuclear	Bechtel	5	6,500		2			3	
Pipelines	Bechtel	8	28,076		3	1			4
(oil/gas)	Fluor	9	25,547	1	2	1			5

F – Feasibility
CE – Consulting engineer
D – Design
Dc – Design consortium
D/CM – Design construction management
DC – Design construct
Source: Kathleen J. Murphy, *Database on Macroprojects 1970–83.*

There is a distinction in award of contracts that can be noticed between the infrastructure projects that is, transportation systems, water supply, urban development, power generation facilities, and the more "industrial" oil and gas pipeline projects. The infrastructure projects award separate contracts for design and engineering, substituting with local construction firms, and government ministries where possible to keep the foreign exchange content down. On the capital-generating pipeline system, contracts are more often let on a design/construct basis.

The Providers of Process Technology

U.S. firms have evolved a highly competent and technologically advanced national petroleum and petrochemical industry. Their technological leadership derived from the pace and nature of national competition provides the basis upon which they have enjoyed international success. It appears that a solid track record and reputation for technology leadership leads to high demand. By the late 1970s, Kellogg had installed 50 percent of the world ammonia capacity, whereas Lummus had installed 51 percent of world ethylene capacity. This confirms that Third-World customers tend to select the leading technologies when making their high-cost technology decisions.

Technology processes are not always installed by the developer of the technology. In fact, design/construct firms who select from and provide numerous technologies have experienced great successes in the macroproject marketplace. In the hydrocarbon processing area, four U.S. companies stand out; these are Bechtel, Fluor, Kellogg, and Foster Wheeler. Although all four provide engineering and design services and are from time to time designated as contractor, more frequently they are awarded project management or turnkey contracts, which they execute on an independent basis (Table 13.2). When compared with European or Japanese providers of the same technologies, these project management contracts are unique to U.S. firms, whereas Europeans and Japanese are more likely to offer fixed-fee packages via a consortium.

The Managers of Complexity

Some of the largest and most complicated projects in the Third World were conceived in Saudi Arabia during the mid- to late 1970s. And all of these phenomenal dreams of enormous new industrial complexes and new cities were awarded to U.S. design/construct firms. The Jubail

TABLE 13.2
LEADING PROVIDERS OF PROCESS TECHNOLOGY: PORTFOLIO

	Number	$MM	Eng/Design	Project Management/ Turnkey	Contractor
Oil refineries					
Foster Wheeler	9	10,390	3	4	2
Fluor	8	5,920	1	4	3
Bechtel	4	2,430	2	2	
Kellogg	10	2,010	3	1	6
Gas proc/LNG					
Bechtel	7	5,824	2	3	2
Kellogg	3	5,100		2	1
Fluor	8	4,561	3	5	
Foster Wheeler	2	1,900	1	1	
Petrochemicals					
Fluor	6	7,180		3	3
Foster Wheeler	2	3,000	2		
Parsons, R. M.	2	1,300	1		1
Fertilizer					
Kellogg	28	6,644	10	12	6
Foster Wheeler	7	1,460		4	3
Fluor	3	996		3	

Source: Kathleen J. Murphy, *Database on Macroprojects 1970–83.*

Industrial Complex, $10 billion, was awarded to Bechtel, the construction manager and master planner. The Yanbu Industrial Center, $10 billion, was awarded to R.M. Parsons, with similar responsibilities. Similarly, the Ruweis Industrial Complex in Abu Dhabi, also $10 billion, was awarded to Fluor. These projects are only the largest of a long list of contracts won by these project management leaders during the last decade (Figure 13.1).

In addition to these firms' ability to build up their organizations to meet such large-scale challenges, they have also been extremely aggressive in marketing their capabilities. During the 1970s, it was remarked that Bob Fluor and Steve Bechtel personally traveled to these developing areas to suggest development possibilities and negotiate a project concept, and a contractual role. The mix of contracts won during the decade demonstrates the complex mix of technologies, engineering, and managerial expertise at the disposal of these top managers to offer in support of these newly rich oil producers.

FIGURE 13.1
MAJOR E/C FIRMS: TRENDS IN CONTRACT AWARDS (macroprojects 1970–83 only)

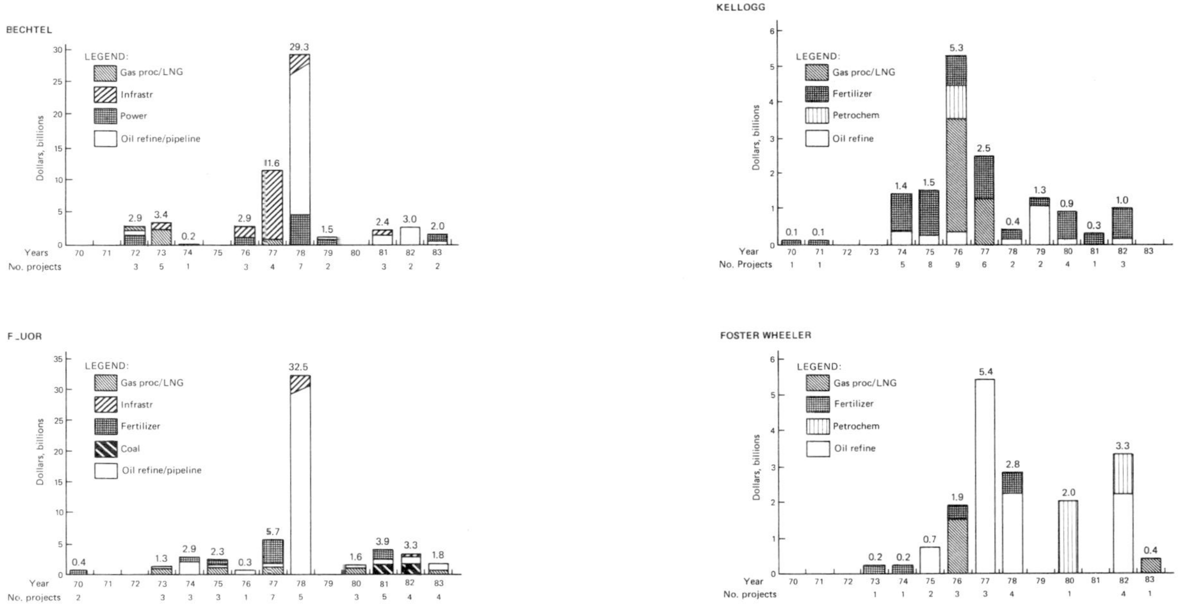

Source: Kathleen J. Murphy, *Database on Macroprojects 1970–83.*

CHANGES UNDERWAY IN THE MACROPROJECTS MARKETPLACE

The demand for engineering/construction services is down. The pace of new project announcements has slowed, and all concerned are searching for the next opportunity to get involved in a major project. Many are looking to the early planning stage for hints of new projects to come.

In fact, there appears to be a change occurring in the development priorities of Third-World countries. To compete effectively, it is important to understand the subtle changes that are taking place.

Redefinition of Development Priorities

Comparing the macroproject portfolio for developing countries for the periods before (1970–73) and after (1980–83) the oil crisis, it is possible to see that the priority development sectors have changed. During the early 1970s, metal extraction and processing were an important and large percentage of the total project agenda. Since 1980, the fascination with metal development as a means of raising foreign exchange has been replaced by a preoccupation with energy sources and substitutes to fuel their economies. Oil refineries, oil and gas pipelines, and coal and synfuel projects are enjoying significantly more interest than the metal or materials processing sector (Figure 13.2).

Of great interest is the large number of infrastructure projects proposed, most of them in the planning stage, and on a scale far beyond what was imaginable a decade ago. The increase is not due to an escalation in project costs, as all costs for project industry segments have moved into higher investment categories over the course of the decade. In fact, the scope of projects being proposed are considerably more complex. Roads are cross country; water supply and sewer systems are citywide; irrigation systems touch entire regions.

One explanation for this large infrastructure project portfolio is the fact that Third-World countries were expecting the value of oil to remain high for decades to come. By the end of the 1970s, it was already evident that oil dollars were being cycled into infrastructure segments of the economy; a roll-over effect was occurring. Developments during the 1980–83 period confirm this trend (Figure 13.3). A high proportion of oil refinery projects were already in the contract let stage, while many of these were postponed or canceled during 1983. In comparison, the number of infrastructure projects that have reached the contract let stage

FIGURE 13.2
MACROPROJECT PORTFOLIO: PRE- AND POST-OIL CRISIS: A COMPARISON

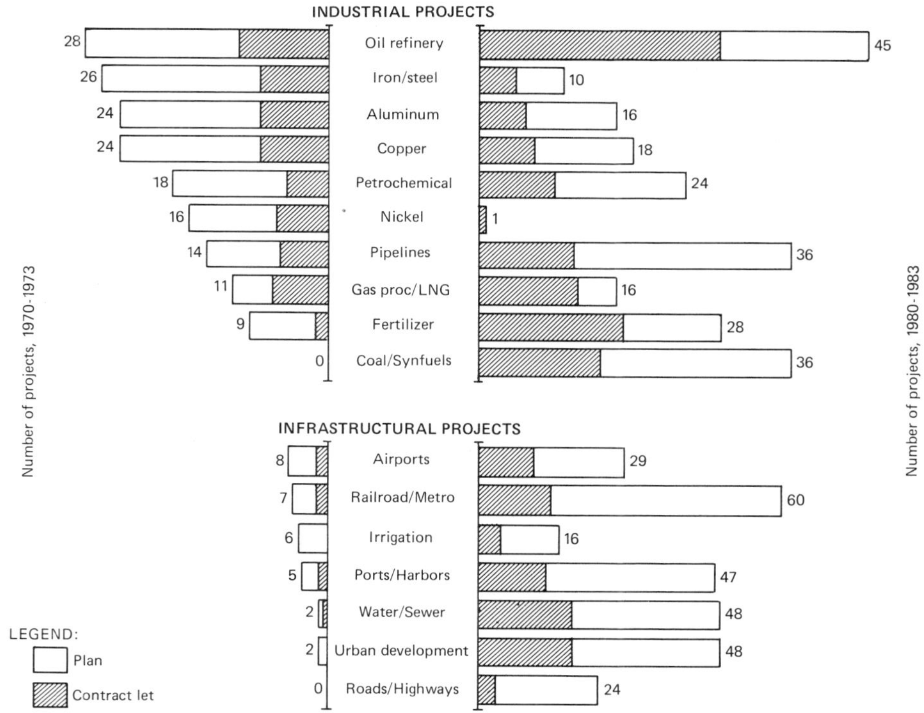

Source: Kathleen J. Murphy, *Database on Mccroprojects 1970–83.*

FIGURE 13.3
THIRD-WORLD MACROPROJECT PRIORITIES: 1980–83

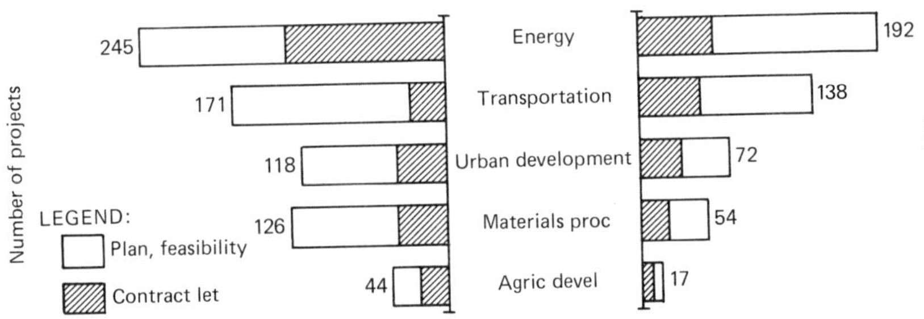

Source: Kathleen J. Murphy, *Database on Macroprojects 1970–83.*

is actually quite low. The list of infrastructure projects might be more significant as an indication of what might have been, should oil prices have remained high, and foreign debt levels remained tolerable, due to an inflow of oil dollars to oil-producing economies.

Grouping the recent portfolio of projects by their end use, the priorities for the early 1980s seem clear: Energy sources and substitutes, including oil refineries, gas processing, LNG, power generation, coal, and synfuels, are attracting the major portions of Third-World planning and investment. This is followed by a strong interest in the development of transportation systems and networks. Both energy and transportation hold the potential for positively supporting the growth of larger segments of these developing economies than their previous heavy interest in metal mining and processing.

Contraction of Macroprojects Market and Customer Demands

Segments of the marketplace have become closed to development over the last decade. Some countries have been closed to U.S. participation, such as Iran and Afghanistan. Others are not in the condition to "construct," i.e., Lebanon and Central America.

There appears to be a state of global overcapacity that is causing a drop in demand for metal products and petrochemicals. Third-World enthusiasm for some of the latest technologies, including nuclear energy and synfuels appears to be waning. There is certainly little reluctance to postpone or cancel a project, should the technology prove to be uneconomical or the market price for the output uncompetitive. Approximately $10 billion in nuclear power plants have been suspended in the last three years, as well as billions of dollars of petrochemical, synfuels, coal mines, and gas processing projects. This contraction in demand is directly related to the economic feasibility of these projects vis-à-vis the price of oil.

In order to diminish the amount of foreign exchange required to implement a macroproject, efforts are being taken to reduce the size of contracts let to foreign companies. Some countries require that local partners be included in a joint venture. Others break the total project down into small packages, many of which become manageable to local companies as they become smaller in scope.

Furthermore, local engineers and constructors have gained a lot of experience from the macroprojects implemented during the 1970s. They are proving to be fully competent at assuming full independent responsibility for many aspects of planning, design, and construction of industrial and infrastructural projects.

TECHNOLOGY VENTURING AS A POLICY RESPONSE

Full management responsibility for the skillful design and engineering of highly complex projects, or the sales and installation of state-of-the-art technology appears to be the only turf remaining for U.S. companies to compete on.

In fact, technology is not the only dimension along which companies can gain a competitive advantage. Macroprojects require capital and market access, in addition to access to technology. The bargaining position of the project sponsor and the supplier is determined by what the project needs and how able the sponsor is in fulfilling those requirements independently (Figure 13.4). Comparing the 1970s with the early 1980s, it is clear that the boom was due to the sudden availability of capital due to the recycling of oil dollars. Western countries were eager to recoup currency losses to OPEC countries through technology transfer. At the same time, there was a high demand for petroleum, petroleum products, and downstream items.

FIGURE 13.4
BARGAINING DYNAMICS AND PROJECT REQUIREMENTS

Leverage Equation	Capital Sourcing	Technology Transfer	Market Access
Low independence, high joint	Insufficient capital	New industry	No local market
	Unpredictable cost estimates	Exclusive technology inaccessible	No access to foreign markets
High independence, low joint	Sufficient capital	Developed local industry	International market channels
	Predictable cost estimates	Local capital goods production capability	High demand assured

Source: Kathleen J. Murphy, *Macroproject Development in the Third World* (Boulder, Colorado: Westview Press 1983).

In contrast, the current structural block in the development process runs along the same dimensions: The high debt levels and decline in the value of oil are making it difficult for certain countries to move ahead with new projects. Over capacity in many segments confirms that there will be no market access for the output once the project is started up; thus, leading to numerous cancellations and postponements.

Dynamics of the Project Bargaining Process

As development efforts become constrained or blocked, there is greater need for collaboration in seeking constructive and creative solutions. Table 13.3 presents three macroprojects that were experiencing serious blocks. The first, a Pakistan highway to cover 1,000 miles, could not get off the ground because of lack of available public capital to invest in the project. The government is now soliciting collaboration with a prospective contractor. It has recently issued a request for proposal that requires that the contractor set up a company to collect user fees during the 10 years of construction work. Numerous financial and tax benefits will be offered, in addition to a government guarantee of the profits.

The second example describes two attempts by Bechtel to promote the feasibility of a trans-Mediterranean gas pipeline that would reach 6,500-feet depths during the sea crossing. The conceptual study, conducted in-house, and submitted to the Nigerian government, contained

TABLE 13.3
PROJECT REQUIREMENTS: PROBLEMS AND SOLUTIONS

Project Requirements	Project	Problem	Solutions (Attempts)
Capital sourcing	Karachi-Nowshera Highway, Pakistan, 1,000 miles, $730MM 1983	Insufficient financing, must come from private sources	Requested proposal: Financial plans using tolls or user fees to recover costs over ten years work Profits guaranteed by government Winning contractor should form an operating company to raise construction funds Waiver of import duties and tax relief
Technology transfer	Nigeria to Europe Gas Pipeline $9,000MM 2,480 miles 1982	Maintenance and repair of pipeline at 6500-feet depth	Bechtel submitted in-house conceptual study to Nigerian government, including three routes (Algeria, Tunisia, and Morocco)
		Political uncertainty	
	Algeria-France via Spain Gas Pipeline, $6,000MM, 807 miles 1983	No financial or market problems, only how to repair pipeline at 6500-feet depth	Bechtel conducting feasibility with Sonatrach, Algeria, and Enagas, Spain
Market access	Chirique Grande Coal Gasif/ Methanol, Panama, $4,500MM 1982	Use midwest U.S. coal	Ebasco leader, forming consortia organizing U.S. coal suppliers and consortia of domestic and international technical groups, equipment suppliers and investors to fund and implement U.S. AID grant

Source: Kathleen J. Murphy, *Database Macroprojects 1970–83.*

three potential routes, but all facing the same problem: How to repair a pipeline at 6500-feet depths. The subsequent Algeria to France via Spain gas pipeline had adequate financing and secure markets for the output of the pipeline; however, it faced the same technical problem of pipeline repair at great depths. A joint feasibility, conducted by Bechtel with Sonatrach of Algeria and Enagas of Spain, is currently underway.

Finally, the Chirique Grande Coal Gasification plant has recently been proposed as an approach to providing an energy source to Panama, while at the same time providing a market outlet for U.S. coal. U.S. coal producers are eager to get this project underway; however, there are numerous impediments to planning and implementing the project. Ebasco has taken the lead and is forming a consortium that includes U.S. coal suppliers, in addition to a consortium of domestic and international technical groups and equipment suppliers. U.S. AID has provided a grant of $500,000 to get the initial study going.

In summary, impossible limitations seem to be addressed by collaborative efforts on the part of interested parties. Creative formulas for project development can be developed, if the right mix of parties is included in the brainstorming process.

Collaboration and Synergy

As prospective partners come together, the opportunity to joint venture allows new dimensions of expertise to be quickly assembled, with minimal costs to the project sponsor and with reduced risks to all. The decision to form a consortium or joint venture is determined by the specific needs and requirements of the project and the capacity of the prospective participants to fulfill the project requirements. Beyond this, the mix of skills and competencies among the partners chosen to participate in the joint venture permit important synergies to occur.

The consortium or joint venture formed to develop major airport projects, detailed in Table 13.4, provided distinctive benefits to the participants as well as to the project. The joint venture between Bechtel and Hellmuth, Okata & Kassabaum on the King Khalid International Airport in Saudi Arabia allowed the design team to achieve the capability to design at the level of scale required — it is the largest in the world — and at the same time permit effective supervision of the more than 59 separate bid packages offered for the construction phase.

Similarly, the Negev Airbase Contractors consortium permitted integration of the design and construction process. The consortium was

TABLE 13.4
THE JOINT VENTURE/CONSORTIUM AND SYNERGY

Synergy Required	Project	Consortia
Managerial	King Khalid International Airport, Saudi Arabia $3,200MM	*Bechtel*: Master plan, design, construction management *Hellmuth, Okata & Kassabaum*: Design (with Bechtel) *Contracts supervised:* 59 separate bid packages - construction, 26 construction support centers, 23 permanent facilities centers
Consultant/ contractor	*Negev Desert Strategic* Air Bases (Ovda & Eilat), Israel, $1,000MM	*U.S. Army Corp of Engineers*: Project manager *Negev air base contractors*: Consortium to design and build Perini Corp., lead Air Base Contractors Guy F. Atkinson TAMS (subcontractor and design) Imported labor from U.K., Thailand, and Philippines *U.K. government*: $800MM grant
Local/foreign	Aryamehr International Airport, Tehran, Iran, $850MM	*TAMS*: Master plan, in a design consortium with *Abdul Aziz Farmanfarmaian of Iran*
Financial backing	Cengkareng/Jakarta International Airport, Indonesia, $536MM	*French government*: $220MM "tied" soft loan and export credits in support of the French consortium: Aeroport de Paris Three others

Source: Kathleen J. Murphy, *Database Macroprojects 1970–83*

composed of designers, engineers, constructors, as well as imported labor under the supervision of the consortium. Their presentation of their proposal as a team that provides coordination of all aspects of the development process enhanced their attractiveness to the project sponsor.

It is more and more common that an engineering or construction company might be required to make a bid with a local partner. In fact, this offers the possibility of gaining access to local practices and unique conditions, thus, offering a final product that is more in line with local requirements.

Finally, many consortia are formed to gain access to government financing, and soft loans, that are available to national exporters of goods or services. In the case described, the French government provided $220

Potential of a National Technology Agenda

To change the course of events, the source of the blockage must be removed. To develop infrastructure, for example, financial limitations might need to be removed — perhaps by barter? To develop large-scale industry, the lack of market demand must be circumvented — perhaps by introducing a new technology?

Any decision to concentrate upon the development of a new technology, whether at the corporate or national level, has worldwide significance. Any new technology can be assured of a demand base around the world, if, and only if, the economics are right.

A case in point is the French government's 14-year investment in the technology development of the high speed train from Paris to Lyon, their state-of-the-art railroad, that is now the fastest train on earth. With government backing, and nationwide collaboration, their end product has recently been launched. This achievement adds to their already established reputation for an efficient and effective mass transit system. Most significant to this discussion is the impact of their global competitive position. Between 1980 and 1983, French companies or governmental bodies and consortia have won nine contracts for rail or mass transit projects in the developing world totaling $4,191 million in project investment. These projects account for almost half of all contracts let, representing three-fourths of total value. The companies involved are listed in Table 13.5.

Whether due to direct exposure and/or participation in the research and development project, or whether due to an improved national reputation for transportation technology, the positive impact of technical breakthroughs achieved appear to be appreciated on a worldwide basis. In contrast, it is difficult to find a comparable level of national collaboration within the U.S. It has not been since NASA's 1960s program that a national-level injection of new technology was experienced by U.S. firms.

In summary, the international marketplace should be considered an extension of the national marketplace. Realizing this, government officials should be more mindful of the impact of locally developed technology on the global competitive position of U.S. firms. Any opportunity to wrestle with technological boundaries at the local level can provide important competitive advantages to national companies competing in the global project development arena.

The United States has been the last to consider forming consortia, in spite of the fact that individual engineering/construction firms have been participating in technology ventures consistently, and most often with foreign firms.

TABLE 13.5
FRENCH PARTICIPANTS ON RAILROAD/MASS TRANSIT PROJECTS

Name of Organization	Number of Projects	Feasibility/ Engineering Design	Consort Member	Other Partners
Sofrerail	3	2		(Delayed)
Sofretu	3	3		
Alsthom-Atlantique	3		1	2 equipment suppliers
Creusot-Loire	1			1 equipment supplier
Bur Centr d'Etudes pour les Equip d Outre	2	2		
Francorail-MTE	1		1	+ Financers
Soc Genl de Enterprise	1		1	+ 17 members (French)
Cia Intl pour li Dev d' Infrastructure	1		1	+ 19 members (French)
Spie Batignolles	1			Civil Engineering
LPEE	1		1	(Delayed)

Note: Total number of projects, 9; total investment, $4,191MM 1980–83 only.
Source: Kathleen J. Murphy, *Database Macroprojects 1970–83.*

We might fear a consortium or joint venture posture in international markets, but perhaps an international consortium with shared financial backing by OECD governments would be more acceptable. U.S. consortia of smaller companies might be less threatening than a consortia of our major firms.

A lot of good has been achieved by the development efforts of the 1970s, even though it looks at times as if the tab has been left to the Western world to pay ($600 billion debt outstanding is equal to the $600 billion of macroprojects that were underway during the 1970s).

The speed with which new technologies will need to be developed and the risk of investing in the wrong technology with highly variable economics warrants consideration of venturing as a serious option for U.S., OECD, and Third-World organizations alike.

14

Public Works and Their Impacts on Infrastructure and the National Economy

Brigadier General Robert J. Dacey

Public Works are the backbone of our infrastructure, our nations life support systems. The Corps of Engineers has a long history of building and maintaining structures that are part of the broad American infrastructure. This includes community services, communications and transportation facilities, and military installations. In its Civil Works activities, the corps develops and maintains structures for water development, management and control. The total construction cost of these projects, built over the last century, is estimated at over $34 billion for the structures, facilities, equipment, and land. But their replacement value would exceed that amount many times over.[1]

It would be difficult, if not impossible, to provide the money to replace this extensive public works infrastructure. We must repair and modernize deteriorated, inadequate, and obsolete structures, as well as build some new facilities to satisfy present demands.

The impact of new construction and rehabilitation efforts by the Corps of Engineers on the nation's infrastructure centers around one of the most critical of our national resources — water. Not only is water essential to all aspects of living — industrial development, energy production, agribusiness, and so on — but it is also an essential element in the development and maintenance of our infrastructure.

THE IMPORTANCE OF WATER ON THE ECONOMY

In his book *Centennial*, author James Michener traces the social and

[1]Major General John F. Wall, Jr., "Resolving Infrastructure Problems: A Challenge to the Corps of Engineers," *The Military Engineer* (July-August 1983).

economic development of the region around the Platte River. The river, he writes, "was the brawling, undisciplined, violent artery of life and always would be."[2] Today, water is still a critical resource in Denver and an artery of life worldwide.

We all understand the importance of water for drinking, bathing, cleaning, and other obvious uses, but the hidden costs of water are present in commodities we use daily. A *Newsweek* special report offered several incredulous examples: Beginning with the water that irrigated the corn that was fed to the steer, a steak may have accounted for 3,500 gallons. The water that goes into a 1,000-pound steer would float a destroyer. It takes 14,935 gallons of water to grow a bushel of wheat, 60,000 gallons to produce a ton of steel, and 120 gallons to put a single egg on the breakfast table.[3]

In 1980, water use reached some 2,000 gallons a day for every man, woman, and child in the United States, a 22 percent increase from 1970. Americans use three times as much water per capita as the Japanese.[4] Of the 50 states, the water use of Texas is second only to California. The two largest water-consuming sectors of the economy are industry and agriculture, which together account for more than 90 percent of the daily total. Within those sectors, power generation and irrigation are the major end users. Public water supplies represent most of the balance of consumption.[5]

A water shortage in the nation's farmlands could be economically disastrous, resulting in food shortages and sharply increased food prices and in declines in agricultural exports, further worsening the nation's trade position. Ripple effects would be felt through the rest of the economy, affecting the entire chain of food distribution and production.

THE IMPORTANCE OF WATER IN THE DEVELOPMENT AND MAINTENANCE OF OUR INFRASTRUCTURE

A great amount of the bulk commodities we rely on are transported on our waterways. One 15-barge tow carries the same cargo capacity as 900 trucks (22,500 tons),[6] making waterborne transportation the most energy-efficient mode for line hauls.

---

[2]James A. Michener, *Centennial* (New York: Random House, 1974), p. 38.

[3]"The Browning of America," *Newsweek* (February 23, 1981), p. 27.

[4]"War over Water: Crisis of the '80's," *U.S. News and World Report* (October 31, 1983), p. 11.

[5]"The Approaching Water Supply Crisis," *Nation's Business* (August 1983), p. 17.

[6]Iowa Department of Transportation.

In 1981, our inland waterways carried about 500 million tons of intercity (domestic barge traffic) freight. Our seaports handled 1.3 trillion tons of cargo that same year. And here we have two serious situations for revamping a portion of our infrastructure: waterways and ports.

The locks that support waterway traffic are old and crumbling. Although the average age of our 184 locks is 40 years, we have 56 locks that are older than their design life of 50 years, including some that are pushing 80 years. They and their gates and machinery are worn from thousands of openings and closings; lock approaches are battered; lock walls are eroded.

Age isn't the only problem. Waterway traffic has increased beyond the range the locks were designed to handle. The result is congestion. Barge tows often must wait eight hours or more before locking through.

The National Waterways Study conducted by the corps to assess the capability of the existing system and to define the nation's future needs found that an investment of $5–12 billion will be needed by the year 2003 to maintain and improve the nation's existing waterway system. The importance of our locks to the efficient movement of goods makes it imperative that a means of funding lock rehabilitation be found.

A General Accounting Office study called the need to deepen our ports the most serious navigation issue before Congress. Our National Waterways Study reached the conclusion that for our nation to remain competitive in world trade we must deepen our seaports to accommodate deep-draft shipping. On our east coast and our Gulf coast not one port is deep enough to handle the large vessels that carry twice the standard shipload.

Clearly we need to address this issue. At the present time there are 21 ports asking to have their channels deepened to depths of 50–55 feet. It is time for us to consider our total national requirements for deep ports including how many we need and where they should be.

When the planned deepening of the Port of Brownsville, Texas, to 55 feet is completed, officials estimate tonnage handled through the port will increase from the 3.5 million tons in 1980 to 19.7 million tons in 1985. Their projection for 1990 is 43.3 million tons. Although the permit to complete this work has been granted, officials are still working on funding arrangements with industry.

Ports and waterways are not only important for our economy, they are also critical for our national defense. In the event of mobilization, waterways would move personnel, supplies, and raw resources to feed the industrial infrastructure in support of the defense effort.

Navigation and defense preparedness are only two sides of this important resource. Another aspect of water infrastructure is the water retention structures and the complex system of water services, pumps, pipes, purification plants, storage areas, and other facilities that move water from its source in farms, factories, offices, and homes of users.

Responding to these needs, the corps has constructed more than 600 multipurpose reservoirs for flood control, water quality, fish and wildlife enhancement, municipal and industrial water supply, hydropower, navigation, recreation, and some irrigation, including 21 reservoirs in Texas.

The Texas 2000 Commission Report, completed by the governor's office in June 1981, reports that Texas has 27 federally built, owned, and operated reservoirs. These federal reservoirs contain 5 trillion gallons (15.4 million acre-feet) of water, almost half the state's total surface storage capacity in its 169 major (5,000 or more acre-feet) reservoirs.[7]

Even with the completion of 13 reservoirs presently under construction and the building of 19 authorized reservoirs, the dependable yield including ground water sources will be almost 6 million acre-feet short of the 27 million acre-feet needed to sustain the socioeconomic structure in Texas by the year 2000. By the year 2030, the deficit is estimated to exceed 18 million acre-feet.

Availability of supply is a nationwide problem. It stems from the misalignment of water supplies and water users. Since more than 75 percent of the population and most manufacturing activity are concentrated on less than 2 percent of the land, the distribution infrastructure is vital. Across the country it, too, is deteriorating.

Most of the country's aging industrial cities face the costly task of repairing antiquated, leaky water systems. In Boston, for example, it is estimated that water lines leak 2 gallons for every gallon delivered, but no one knows for sure because much of the city's water flow is unmetered. New York City's water system loses about 100 million gallons a day.[8] (The Southwest, I should add, is nearly 100 percent metered. Dallas officials knew how much water the city lost when pipes burst during the 12 consecutive days of below freezing temperatures in December. That crisis served to remind us of the vital importance of this system.)

---

[7] *Texas Past and Future: A Survey*, Texas 2000 Project (Austin, Texas: Office of the Governor, June 1981, revised March 1982), p. 218.

[8] "Water: Will We Have Enough to Go Around?" *U.S. News & World Report* (June 29, 1981), p. 37.

Wastewater treatment facilities also need rehabilitation. Fewer than one-half of the 3,700 largest U.S. cities can meet federal sewage standards.[9] Rehabilitating these systems will be reflected in taxes at all levels of government.

The installations that support our national defense are not immune. Each military base has its own "public works" infrastructure of roads, buildings, heating plants, sewage disposal systems, and housing that the corps maintains. Like our public works infrastructure, they have deteriorated. Why? The same reason — a lack of funds.

We can use the nation's unemployed to help with infrastructure problems; we have appropriated $389 million in the fiscal year 1983 budget for this purpose. Those Jobs Bill (H.R. 1718) funds are being used on existing construction projects such as for stream-bank protection; small flood control and small navigation; needed repair, restoration, and improvements at corps-operated navigation locks and dams and on flood control reservoirs; and construction and maintenance work on the Mississippi River and tributaries. An additional $73.7 million was appropriated for improvements in the maintenance and repair of Army family housing.

The funds provided by this bill are an excellent start but are only a beginning to the amounts needed for the immense task of rebuilding America.

WATER PROBLEMS

In addition to the four basic aspects of assuring long-term water supply — availability, adequacy of storage and distribution facilities, maintenance of water quality, and conservation — we also have to consider many additional technical and institutional problems.

Populations are shifting to the Sun Belt, many areas are already short on water and economies are changing. The out-migration resulting from the dust bowl in the early 1930s had tremendous social and economic impact. Today, as we consider interbasin transfers — moving water where the people are — we are faced with a myriad of legal, political, environmental, and economic issues.

Uses for water compete with each other. New technologies in the energy field such as coal slurry, coal gasification, and oil shale are all

[9]Ibid, p. 34.

large water consumers. Is irrigation or energy development a more critical use of water? Hydropower or navigation? Should water allocations be optimized when based on purposes or consumption projections? These are questions that must be answered, priorities that must be set.

Pollution from contaminates, such as acid rain, polychlorinated biphenyls, and toxic chemicals, is a major concern. Quality of water is as important as supply. Overall, what is the appropriate emphasis on environment, and who decides?

Conservation practices, too, are a major consideration; but conservation alone merely delays the crisis; conservation will not solve it. Recycling of water has received attention as a solution. Instead of returning the flow to the system, however, reuse means a downstream user loses his water supply. As Burton Crane wrote, "There ain't no such thing as a free lunch."

Forecasting long-term water resource development needs is another difficulty. Our planners must rely to a large extent on past trends to predict future requirement. But the past is not always an appropriate prologue. Events occur that have no precedent. The development of a new technology, a shift in national or regional priorities or a caprice of nature are examples of such occurrences. These can result in changes in industrial development patterns, shifts in population, or changes in per capita income. Each has an impact on water resource needs. Such unforeseen events can significantly distort a future based on an extrapolation of past trends.

NEW TECHNOLOGIES

In solving our infrastructure and water problems, we have an opportunity to find better ways of doing things. The Corps of Engineers has five research and development laboratories and seven division laboratories that in coordination with academic institutions keep us on the cutting edge of technology in our civil and military roles.

We have begun a state-of-the-art research and development program in the construction, operations, and maintenance work areas. Historically, relatively little research effort has been expended on rehabilitation or maintenance work. Our goal now is to develop new techniques and materials that will help reduce costs and improve effectiveness of our maintenance program as well as construction and operations. We have good opportunities for research on repair and rehabilitation of locks and dams, embankments, navigation channels, coastal structures, and electrical and mechanical systems among others.

Other areas, such as aquifer recharge, which is in a primitive state of technical development, still need to be explored by the research and development community. There have been many proposals in aquifer recharge but very few tests of this important opportunity to the Southwest.

The corps' successful accomplishments on special projects such as the Waste Isolation Pilot Program, the Strategic Petroleum Oil Reserve Program, the Pantex project, and the use of remote sensing capitalizes on our laboratories and districts working closely together to develop the latest technology in the areas of planning, design, and construction. The corps, with only 5 percent of the Army research and development (R&D) community, recently captured 22 percent of the Army's R&D achievement awards.

WHY FEDERAL DEVELOPMENT?

As the southwestern division engineer, I serve all or part of eight southwestern states. We try to formulate and operate projects to optimize national growth, which may not always coincide with the preferred direction of the state of Texas. We strive to serve in a nonadvocate role. We are not in competition with states and locals, but instead we work with them in partnership efforts. This approach strives to ensure that designs bring both short- and long-term benefits considering both the region and the nation.

Risk taking, a predominate theme of this volume, is a major consideration. The federal government is a risk underwriter. Large water projects are capital intensive with long-term paybacks. Could private industry, for example, undertake a massive project such as the McClellan-Kerr Navigation System? That project, I must note, produced a realized rate of return in excess of 13 percent of invested capital between 1971 and 1980, an attractive investment compared to private-sector returns. Capital stock produced a return of about 4.7 percent during 1968–78.

Any representative of the several federal agencies involved in the water resource development area could present similar arguments for federal involvement. The programs administered by these agencies have served our nation well in the past and, with prudent levels of investments, will continue to do so in the future.

CONCLUSION

There is no question that large-scale public projects are needed. It is not a question of *whether*, but *when*. Water problems are a matter of crisis resolution. Like the energy crisis, the water crisis can be managed and resolved if the nation is willing to commit the necessary resources.

Considering today's economic realities and priorities, it is obvious that it will require more than federal resources alone to solve this crisis. A critical issue facing the water resource development community is determining on what financial terms federal assistance will be provided. We in the administration are deeply committed to a new partnership with the states and other nonfederal sponsors in the development of their water resources, both in the financing and in the planning, construction, and management of these projects. The administration and Congress are attempting to develop cost-sharing formulas that are equitable and responsive to economic conditions.

Our role, as a federal agency, is to analyze problems, sort out the technical implications, determine ways to prudently underwrite the risk, present alternatives, and advise the decision makers. Engineers planned and built our nation's military and civil infrastructure. It will be engineers who, through their technical expertise, will provide the improved methods, materials, research, and value criteria to solve our national infrastructure problems in the future.

15

Community Planning for Technology

William Gregory

I had some extracurricular activity as a local planner in my earlier life in the New York area. It's a question of state and city competition to attract high-technology industry. I watched this going on as a citizen in my own community in northern Virginia, and I wondered when the planning for the infrastructure, the community facilities, the educational facilities, and the like would ever catch up with the salesmen and their promises and their blindness of the industry site selection committees.

In the Washington area, local planning is beginning to be something of a disaster. We have seen Los Angeles as a case example of an enormous aerospace high-technology industry complex. But industry is beginning to turn away from Los Angeles because of the urban chaos. Hughes has decided to put a new plant in Florida, though it has concentrated most of its facilities in the Los Angeles area. (The urban problem is not the sole reason why Hughes is going to Florida; they had some practical reasons about producing large spacecraft and not wanting to truck them. But it is still the sign that Los Angeles has topped out, and its decline is starting.)

TWR is headed the same direction. It had to open a couple of new facilities late last year; and instead of going to the Los Angeles area, where they are concentrated, they are headed, believe it or not, into Washington D.C.

I see the same planning mistakes, or the lack of planning, that could turn Washington into another Los Angeles a decade or so from now: strip development, inadequate provisions for future traffic, slow progress on rail transit, and especially a lack of coordination between jurisdictions. Traffic engineering in Washington seems to be devoted to designing bottlenecks. It is only in Washington that the federal and local governments could spend a $100 million on a new interstate highway extension into the heart of the city and then try to restrict it to four-person car pools during the morning and evening rush hours.

Governor White of Texas makes a good point when he states that as far ahead as Texas seems to be when compared to the rest of the country, there is still much planning to be done. I've been reading about Houston's problems lately, and I will be curious to see what happens to Austin as it begins to develop.

Major corporations often have no choice as to where they must locate, but I am really surprised that industry hasn't tried to crack a bigger whip in community planning. After all, high technology has to attract good people, and good people demand a pleasant working environment.

16

Industry and Government in Space: Making the Long-Term Commitment

John J. Egan

Remarkable as it seems, the Space Age was ushered in only a quarter of a century ago, in 1957, with the orbiting of the first artificial satellite. Yet, for today's generation, the beginning of space flight has receded into contemporary history, even as America's space shuttle, the European Ariane launch vehicle, permanent Soviet space stations, and exciting prospects for commercial opportunities in space combine to open a Second Space Age of staggering potential.

In this short period of time, a mere third of an average citizen's life span, the United States has landed six two-man crews on the moon and explored with unmanned probes the farthest reaches of the solar system. Since 1957, a multibillion dollar communications and data transfer industry has emerged using satellite interconnects; nations have grown dependent on a worldwide satellite weather observation system; satellite remote sensing has yielded benefits ranging from better crop forecasting to more efficient exploration for mineral and oil deposits, and new industrial processes have been investigated.

The indirect economic benefits ("spin-offs") from space research are almost impossible to calculate. They range from enhanced large-scale project management techniques to energy storage systems and composite materials and give every indication of almost indefinite expansion.

We recognize that the fundamental task of space research, which is the charter granted to NASA, is to expand the envelope of humanity's understanding of the universe. However, our expanded knowledge of the space environment has yielded new concepts and technologies of commercial interest. This chapter addresses the business implications and opportunities resulting from such space activities.

From the very beginning of the Space Age, monies invested in space projects were justified on the basis of improving national security (via military surveillance and communications satellites), contributions to basic scientific knowledge and, occasionally, their long-range economic potential. Military satellites, with their ability to verify data of military importance, have played a strategic role in maintaining the ongoing balance of power between the superpowers, and the scientific community can point with pride to the myriad discoveries involving the Earth, moon, other parts of the solar system, stars and galaxies beyond, and reactions of the human organism to the space environment. These are the scientific rewards of research in space.

There are also practical, down-to-earth benefits, too. Communications is the most evident, widely applied, and lucrative product of space research. A whole new spectrum of actual or potential services includes direct broadcast television, global search and rescue, secure business communications, navigational aids, electronic mail, package locators, and "personal wrist radios." Still other innovative applications will be both technically and economically practical within the next decade.

Yet, with the exception of communications satellites, little in the way of commercial gain has emerged from the space endeavor. That situation, however, is rapidly changing. The private sector is becoming more involved in space technology, once the sole province of government-funded research and development. The displacement of the U. S. government monopoly in space development is heralded by the emergence of private enterprises seriously pursuing ways to benefit commercially from NASA and other federally sponsored work in space transportation, remote sensing, and industrial processes.

THE ROLE OF INTERMEDIARY

It is our assumption at Coopers & Lybrand that these areas will eventually represent the same relatively safe investment opportunities as communications satellites do today. As in the financing of any large commercial project, there is a need for these technologies to become better understood, as well as for the financial risks to be spread among more investors. Once the nonbusiness risks are minimized, the more mature technologies are separated out from the experimental ones, and technical possibilities are clearly evaluated in terms of market opportunities, it is virtually certain that private financial institutions will be increasingly attracted to commercial space projects.

In recent years, we have worked closely with NASA and space-related industry in assisting companies to begin thinking about space. One of our objectives has been to collect information from a number of major corporations concerning how the properties of space may be applied to their business operations. As with any complex business endeavor, the successful development of space depends on a cooperative relationship between users and suppliers (e.g., the space vehicle designers, builders, and launchers). In turn, these ventures must ultimately rely on capital markets and the government for access to the space environment. What makes such institutional networking exceptionally challenging is the fact that space products and services may well reshape the entire marketplace.

To assist corporations and organizations (public and private) in capitalizing on the remarkable commercial potential of space, we have developed a comprehensive set of business planning, marketing, and financial consulting services. Clearly, there is a need for an organization acting to bring together diverse elements in the space field — to serve as business advisor and coordinator. We are committed to fulfilling such a function. By working with a variety of governmental and industrial entities, we have developed in-depth knowledge and understanding of the major participants in the satellite industry, satellite launch services, and other space endeavors.

BUSINESS IN SPACE: A RISKY VENTURE

It is noteworthy that, in 1977, when NASA began its efforts to involve commercial investors in the use of the space environment, they found few who were interested. Now, seven years later, factors such as the successes of the communications satellite industry, the increasingly routine access to space made possible by the space shuttle, and experiments which indicate a possible commercial pay-off from space, have all acted to galvanize industry/investor interest and bring about their active participation in space projects.

Perhaps the most commercially promising prospects in space involve space-based production, especially the application of microgravity in industrial processes. Recent experiments on the European Spacelab (designed and built by the European Space Agency for on-orbit, manned experiments aboard the space shuttle) have demonstrated that the absence of gravity allows the mixing of metals to create light, strong alloys unobtainable on Earth.

This unique characteristic in space has induced McDonnell-Douglas and Ortho Pharmaceutical, a subsidiary of Johnson & Johnson, to develop an experiment for use on the shuttle to study the possibilities for space-based manufacture of pharmaceuticals that, due to the presence of gravity, are difficult to produce in sufficiently pure quantities on Earth. Although this process is still in the research phase, it does suggest the possibility of a potential market for space-produced pharmaceuticals such as urokinase for the treatment of kidney disease and Beta-T for curing diabetes. This is the first example of industry awareness of space as a workplace worthy of serious consideration.

Well-known aerospace corporations, such as General Dynamics, Rockwell International, McDonnell-Douglas, and Grumman, have been joined by other large nonaerospace corporations, such as Johnson & Johnson, Federal Express, John Deere, and others, in an effort to explore the commercial possibilities of space. In addition, new entrepreneurial firms, with names as unconventional as their business plans, have been seeking venture capital to enter the space field. SPARX, Transpace Carriers, AstroTech, Starstruck, Space America, Orbital Sciences Corporation, GeoStar, Orion, and others make up this rapidly growing group.

It would be misleading, however, to imply that the prospects for business in space are without considerable risk. In most areas of business opportunity, extensive private investment is supplemented by government research funds. Quite the opposite holds true in the space business. Until very recently, the private sector has invested relatively little in space-related research and development for the purposes of exploring new business opportunities.

There are many reasons for such limited investment, but two factors are particularly significant: the large initial capital outlay required for access to the space environment and the related issue of undefined markets for space-produced products and services that make it difficult for a business manager to predict a given return on investment. Even more problematic is the fact that, in some cases, although the market is known (e.g., better-quality semiconductor crystals), there is a need for scientific proof that space-based production will yield cheaper, better products than a ground-based industrial process.

Since most of the early efforts in space have been dominated by government, the involvement of private financial institutions has so far been limited. It violates the basic business sense of Wall Street to support the investment of large sums of money in high-risk, long-term space projects, especially when contrasted with conventional, Earth-bound activities.

(It is interesting to note that in Europe banks have traditionally been involved in space projects, such as Ariane, as shareholders and as a source of loan capital.)

However, as space-generated products and services become better defined, entry costs to users will drop and space "as a workplace" will become routinely accepted. As this occurs, financing of space ventures will become easier to arrange. Already, financial institutions have been increasingly willing to fund the purchase of communications satellites and transponders because such satellites are now regarded as relatively safe and lucrative investments. The cost of satellite transponders ($10 to $15 million) is a small part of the overall cost of the satellite, and insurance is available to cover service interruption and business loan. Also, through the use of sale/lease arrangements, the tax benefits that result from transponder ownership can be sold to third parties.

DEVELOPING THE SPACE FRONTIER: PAST ACCOMPLISHMENTS AND NEW OPPORTUNITIES

During the past quarter century, all of these space activities have been made possible by the fact that much has been learned about the distinguishing characteristic of the space environment. The knowledge we now take for granted resulted from the pioneering efforts initiated after Sputnik — from the early U.S. Explorer satellites that penetrated the Earth's radiation belts, in the late 1950s, to the 72 varied experiments carried out by Spacelab's crew last fall.

By pursuing such research, scientists have gathered data on the space environment that is the current basis of industrial interest in this new frontier. No better illustration exists of the integral relationship between so-called pure and applied research. Of particular interest to industry is the stability of the space environment: the near absence of unpredicted disturbances and random forces due to microgravity and a near perfect vacuum; the significance of very small external forces (surface tension, atmospheric drag, gravity-field harmonics, electromagnetic interaction, and radiation pressure) in the long-range behavior of space structures; the fact that the laws of physics apply with equal precision to all bodies, spacecraft as well as planets; the ability to employ digital computers to reliably predict the behavior of complex systems; and the advantages of having a global "bird's eye view" of the planet Earth.

Before describing some private-sector efforts to explore investment opportunities in space, it is worthwhile reviewing the evolution of key space technologies.

Satellite Communications: Novel Idea Leads to Profits

As noted earlier, in the early stages of the Space Age, little thought was given to economic potential. The first U.S. artificial satellite, Explorer 1, was developed only partially for scientific research; it also served as a vital morale booster for the nation. As was the case with so many space projects, which emerged in the wake of Sputnik, the boundary between politics and science was difficult to discern.

However, there were a few pioneers, such as Arthur C. Clarke, who believed that space offered commercially attractive features: a high vantage point above the Earth's atmosphere, microgravity, and a near perfect vacuum. Clarke originated the revolutionary concept of geostationary orbiting communications satellites, i.e., if a satellite were to be placed in an orbit (approximately 22,000 miles) above the equator, where its speed equaled that of the Earth's rotation, it would appear motionless above a single point on the equator. Three such satellites, spaced equally around the equator, could cover the Earth with radio and other electronic signals.

The dramatic impact of communications satellites was first felt in December 1958 with the launch of a primitive satellite communications experiment called SCORE (Signal Communications by Orbiting Relay Equipment). Containing only a tape recorder, receiver, and transmitter, all battery powered, it relayed a message of peace from President Eisenhower on Christmas Day. After 12 days, its batteries went dead.

SCORE was followed two years later by Echo, a 100-foot-diameter aluminum-coated Mylar balloon inflated in orbit. It merely performed as a passive reflector of radio signals beamed to it from Earth. Echo quickly proved impractical — by the time radio signals were bounced back to earth receiving stations, they were barely audible.

After Echo, active communications satellites were developed and launched. With technical advances in rocketry, more powerful launchers became available to place larger, more complex satellites into geostationary orbit. A technological and commercial threshold was crossed in 1962 with the successful launch of American Telephone and Telegraph's Telstar, which transmitted the first live television broadcast from Europe in this country. It was a landmark event: an orbiting facility was funded and operated by private enterprise.

Today, communications satellite services exist for a diverse range of markets: international, maritime, public service, amateur, and domestic, not to mention military satellites that carry 80 percent of all U.S. defense communications.

Given the fact that Echo was only launched in 1960, it is astounding to see how quickly communications satellites have become a commercial enterprise. The interest of the private sector in developing a communications satellite system was demonstrated by AT&T's application to the Federal Communications Commission (FCC) barely two months after the launch of Echo. Recognition of the commercial potential of this new technology led President Kennedy to announce a policy encouraging industry to develop an operational global civilian communications satellite system.

In 1963, the Communications Satellite Corporation (COMSAT) was created by Congress as a private corporation to facilitate Kennedy's directive. Two years later, COMSAT's first project for the International Telecommunications Satellite organization (Intelsat), Intelsat 1, was in geostationary orbit transmitting live television broadcasts between Europe and the United States. Since then, there have been five series of Intelsats spanning the globe, each more sophisticated than the last. By 1983, 105 nations (with COMSAT representing the United States) were members of the system.

Last year, worldwide revenues from communications satellites reached over $1 billion with the cost per year of placing one satellite circuit in orbit dropping from $30,000 for Intelsat 1 to less than $1,000 for the current generation of Intelsats. With the entrance of new ventures, such as Satellite Business Systems, a consortium of COMSAT, Aetna, and International Business Machines, communications satellites are beginning to provide an even greater variety of services.[1]

The first long-range communications service by U.S. domestic satellite carriers was initiated in 1974. The three original carriers, American Satellite Company, RCA-Americom, and Western Union, began service almost simultaneously. By 1982, there were six domestic carriers operating in the United States, with three or four more anticipated to be in operation by 1985. Almost immediately upon initiation of satellite service, U.S. users, including commercial and government customers,

---

[1]Figures quoted from Paul Kinnucan, "Space Business," *High Technology*, October 1983, p. 52.

benefited from a 50 percent reduction in long-distance telephone rates. At the same time, other communications services, such as high-speed data transmission and video broadcast, became widely available at affordable rates.

With more than 50 percent of the U.S. work force involved in the information industry, there is no question that the communications satellite industry is a major and vital segment of the nation's economy. Given the industry's experience in space communications, the question arises, Why haven't entrepreneurs and financial institutions been more eager to invest in other space technologies?

For one thing, a clearly defined market existed for communications services, which is not true for such nascent space ventures as materials processing and remote sensing. For another, the communications industry had over a century in which to evolve a price structure and a network of users, nationally and internationally. Thus, communications satellites, although a new technology, could plug into an already existing infrastructure.

In comparison to some of today's space projects, communications satellites had several technical advantages: simplicity (proper positioning above a specified Earth point and the ability to reflect radio waves to another point on the planet); ground-based research and development resulting in lower research costs; a worldwide demand for voice and high-speed data communications services.

Remote Sensing: Locating the Market

Remote sensing of the Earth from space has proven to be an invaluable approach to acquiring data about our planet's resources. Using special infrared and other sensors, remote sensing satellites have provided key information in areas such as distinguishing healthy from diseased crops, identifying the scope of pest infestation, assisting in forestry planning, detecting air and water pollution, and locating mineral and oil deposits.

The world's first remote sensing satellite was launched by NASA in 1972. Since then, three more of these satellites, designated Landsats (now operated by the National Oceanic and Atmospheric Administration) have been launched. The Landsats performed exceptionally well, although the current satellite (Landsat 4) is in terminal condition. The government intends to orbit a rapid replacement. (At present, there are no plans to launch a follow-up to Landsat 5.)

Over the years, users of Landsat data, which include individual companies, governments, and international institutions, consider it to be an invaluable source of information. However, they have had problems with the way the government's institutional arrangement fails to collect and distribute the information in a timely manner. Congressional testimony is replete with complaints about the program lacking clear direction and suffering from a lack of government commitment to data continuity. To rectify this situation, both the Carter and Reagan administrations encouraged the transferring of land remote sensing to the private sector. In fact, in late 1983, the government sent out requests for proposals for businesses to operate the Landsat system.

Regardless of who operates the system, the users of the data have clearly identified the need for continuity and quality of data, as well as assurance of predictable prices (the government intends to stop making data available for nominal fees). A number of U.S. corporations have indicated a desire to satisfy these needs; however, in sharp contrast to the communications satellite business, the remote sensing market is ill-defined. No space company questions the economic potential of remote sensing, but, given the government operation of Landsat as a public service, no one had the opportunity to define a market price for this data.

Although users acknowledge the difficulty of assigning a meaningful price tag to remotely sensed data, they are convinced of their need and argue that the present system's capabilities require both simplification and improvement. Some American entrepreneurs are considering a low-cost remote sensing satellite system that would accomplish both of these objectives. Meanwhile, France, using technology developed in the United States, has developed a remote sensing satellite (the SPOT system) and they have already initiated an intense marketing campaign to attract U.S. customers. More recently, a new U.S. corporation (a joint venture of COMSAT, Messerschmitt-Boelkow-Blohm of Munich, and the Stenbeck Assurance Company of New York) has announced its intention of using remote sensing technology on the German-developed reusable Shuttle Pallet Satellite System (SPAS).

Space Manufacturing: Greatest Profit Potential

The idea of manufacturing in space is at once intriguing and risky, but this area represents the most lucrative and potentially expansive area of space-related commercial enterprise. NASA and numerous corporations are interested in employing the unique space environment for the production of pharmaceuticals, semiconductor materials, optical glasses, electronics, ceramics, magnets, industrial tools, and special alloys. They are also interested in carrying out space-based research that in time may lead to unforeseen products and services.

The virtual lack of gravity in space eliminates convection currents, allows the mixing of immiscibles (i.e., substances that do not normally mix), and greatly reduces fractures in crystal growth. The vacuum of space permits greater purity and quality control by reducing unwanted gases present in the best commercial vacuum pumps. Recent experiments on Spacelab demonstrate that the absence of convection currents permits the growth of purer crystals in space from which superior electronic chips can be made.

The concept of materials processing was developed early in the space program. Engineers studied the behavior of propellants in rocket stages, while others worried about the flow of metals if the space structures were welded together in space. Meanwhile, drop towers and aircraft flying short parabolic trajectories allowed scientists to evaluate the mechanics of microgravity for several seconds at a time. (John Deere & Company has been using high altitude aircraft to study the impact of microgravity on carbon particles in cast iron.)

The earliest on-orbit biological experiments were conducted by the Soviets on a Vostok spacecraft carrying tissues and microorganism cultures. Our Biosatellite 2 program indicated that improved drugs and hormones might be available in microgravity. Apollos 14, 16, and 17 performed several elementary microgravity experiments, including composite casting, electrophoresis (use of electric fields to separate out fluid mixtures), and the transfer of fluids in weightless conditions.

The most dramatic results came from materials processing experiments aboard Skylab, consuming some 160 hours of crew time. Many of us recall the films of the astronauts performing acrobatics and colored drops of water (grape- and orange-flavored drink) colliding without losing their distinctive coloring. This same principle was demonstrated on the recent Spacelab mission in which experimenters mixed aluminum and zinc, metals that together make a very strong and light alloy that could be used in the future to build space structures. On Earth these metals do not mix due to their different melting temperatures. But, in space, the two make a uniform, strong mix. Much more information about materials processing will result from experiments planned for the space shuttle/Spacelab over the next few years.

Despite the potential of space processing, the nascent state of this technology and the science on which it is based, together with the lack of clearly defined markets, place this endeavor in a high-cost, high-risk category beyond the financial threshold of most commercial concerns. Although a number of firms perceive a definite need for better-quality, space-produced crystals (e.g., gallium arsenide for use in infrared sensors and high-speed computers), the responsibility for demonstrating the scientific, technical, and economic feasibility of space processing will remain with the government acting independently or in joint ventures with business. The policy challenge for government and industry is deciding at what point a concept has been sufficiently developed for government to be removed entirely.

Launch Vehicles: Growing Commercial Interest

The commercialization of space transportation is another area of increasing interest to industry. Proposals for commercialization currently focus on three alternatives: (1) commercialization of U.S. expendable launch vehicles (ELVs), which, under current policy, are being replaced by the space shuttle; (2) transfer of shuttle operations to the private sector; and (3) private-sector development of new upper stages (to boost payloads from the space shuttle's low orbit to higher orbits) and low-cost launch systems.

Originally developed with government funding, existing ELVs, with names familiar to space enthusiasts, such as Delta, Atlas/Centaur, Titan, and Scout, were the veritable workhorses of the space program until the advent of the space shuttle. Once the government committed itself to the reusable shuttle as the primary space transportation system, the private sector has shown growing intest in commercializing some of these veteran vehicles.

The companies involved, as well as those seeking to design their own low-cost rockets, are convinced that the projected growth in demand for communications satellites (General Dynamics anticipates a need for 245 satellites between 1986 and 1995, with potential revenues of $10 billion) warrants the existence of vehicles that complement the shuttle. The recognition of such a market has led to the establishment of Arianespace (a European, primarily French, company), which, with its own expendable rocket, the Ariane, is aggressively competing for shuttle customers. (Arianespace officials see a potential market of $30 million per satellite between 1985 and 1991.)

Companies interested in commercially producing the ELVs (Transpace Carriers for the Delta and General Dynamics for the

Atlas/Centaur) will have to assume the marketing and operations tasks from the NASA in order to be successful. In line with Reagan administration policy, the space agency is working closely with industry to facilitate an orderly transition from exclusive government control of launch facilities to commercial operations. Such transitional issues are complex. Unless a company plans to build its own launch facility (several are proposing to do just that), it will have to make arrangements for lease or purchase of the government facilities.

The administration recently directed the Department of Transportation to take the lead in arranging an expeditious licensing process for commercial launchers. Firms seeking to commercialize launch vehicles anticipate the development of a regulatory framework balancing the government's need to protect the public and national security with the benefit to the nation of encouraging a potentially profitable new industrial sector.

THE "GOLDEN SPIKE" IN SPACE

In his remarks, welcoming home the crew of the space shuttle Columbia, on July 4, 1982, President Reagan suggested a fitting metaphor to describe the new operational status of the shuttle program:

> The fourth (and final test flight) of the Columbia is the historical equivalent to the driving of the golden spike which completed the first transcontinental railroad. . . . The test flights are over, the groundwork has been laid, and now we will move forward to capitalize on the tremendous potential offered by the ultimate frontier of space.

This metaphor is an appropriate description of NASA's current efforts to develop the space shuttle into a space transportation *system*, the transcontinental railroad of the new Space Age. Just as the railroad played a pivotal role in opening up the American West, the goal of regular shuttle operations is to open space to industrial activity on an unparalleled scale. Thus, NASA's shuttle planners aggressively seeking to increase their annual flights, from the 11 planned in 1984 to more than 20 a year planned by the end of the decade.

Routine access to the space environment is crucial to the future commercial success of the space technologies discussed earlier. The space shuttle is a critical step in that direction. It provides capabilities to use and explore space that are not available with expendable launch vehicles alone.

The concept of reusable, reliable, and affordable transportation to space was discussed as early as the 1940s. Technological developments in the wake of the Apollo program greatly enhanced NASA's ability to meet the complex challenge of developing high-performance rocket engines and the ability to control a winged vehicle flying at hypersonic speeds in both space and the atmosphere.

In addition to carrying a larger payload to low earth orbit (65,000 pounds), the shuttle offers unique features including: manned and unmanned on-orbit experiment opportunities, payload deployment, servicing and retrieval of satellites for maintenance or replacement, and, with the Spacelab, a shirtsleeve environment for research activities.

From a business standpoint, the most important feature of the shuttle is its reusability. The more the shuttle is seen routinely placing payloads into orbit, the more exposure scientists and engineers are given to working in space, the more informed business is about space as a serious investment opportunity, and the more space-based opportunities will be seen in standard business terms. Put another way, declining attention by the media to space events means that this new frontier is being incorporated into society as an everyday occurrence. Given Wall Street's reluctance to support anything exotic, less sensationalist press attention is very welcome news indeed.

The shuttle represents much more than another launch vehicle. In combination with the European Spacelab, it can be a flexible, manned space platform for research and development, with the potential of evolving into an enormously capable space system. Already industry is seeking to take advantage of the shuttle's commercial potential.

Earlier, we made reference to McDonnell-Douglas experiments in the space-based processing of pharmaceuticals aboard the shuttle. In 1984, a significant event in space commercialization was achieved when McDonnell-Douglas decided to send into space the first non-astronaut-trained crew member, an engineer named Charles D. Walker, to monitor the progress of his company's experiments. Nothing better illustrates the transition from exploratory venture to routine operations in this new frontier than the conclusion by a company that it is both safe and beneficial to place an employee on such a mission.

In order for business to exploit the newly emerging in-space technologies, it is absolutely essential that the nation continue to develop the capabilities of the space transportation system. If the shuttle is to carry the implications of the president's metaphor of the "golden spike and the railroad" to its logical conclusion, one would expect to see the system's capabilities expanded to meet the diverse needs of the public and private sectors. Such expansion would include increasing the shuttle's stay time in orbit (that period is currently limited to ten days), the development of upper stages to launch the heavier spacecraft emerging from commercial spacecraft programs, and on-orbit facilities to provide for continuous habitation in space.

A permanent space station should be the next major NASA venture. Given a new start in fiscal year 1985, a small four- to six-person facility could be in orbit by 1992 at an estimated cost of $8 to 10 billion. Designed on a modular basis, it could be added to as required. With sufficient space-generated revenues, it is conceivable that industry will eventually pick up where the government leaves off by leasing or purchasing government-funded facilities in space or erecting their own commercial factories.

The operational status of the shuttle, combined with the presence of technicians in space to check out payloads prior to deployment (or to use a screwdriver to repair an errant data recorder as happened on the first Spacelab mission), which greatly lowers the capital investment in performance-assurance equipment, has acted to awaken private-sector interest in the commercial promise of space.

MOVING FORWARD WITH SPACE COMMERCIALIZATION

Until now, our discussion has focused on the unique properties of the space environment and the new products and services that may result from space enterprise. There is an unfortunate tendency, however, for people to perceive space activities as undifferentiated, each offering equal opportunities.

As with any major investment decision, there are vital considerations regarding the sophistication and reliability of a given space technology and the existence of a market for the product. Unless this distinction is made, confusion will arise between technologies that merely offer opportunities for research and those with real commercial promise.

In the exploitation of any frontier, ideas for reaping profits are rife. Space is no exception. Every day seems to bring yet another proposal, sometimes from fledgling entrepreneurial firms, sometimes from major corporations, to develop some innovative space system — be it a new low-cost rocket, a novel space manufacturing process, or mining rare metals from the asteriods.

Given our interaction with the leading actors in the space field, and with the financial institutions and government organizations being approached to fund various space projects, we at Coopers & Lybrand are continually faced with the challenge of evaluating the merits of a given proposal. The criteria one uses for discerning the "far-out" notion from the one with real commercial possibilities are similar to those applied to any large project. It is a matter of assessing the project's marketability and return on investment against the costs and risks of undertaking the project. However, in comparing space ventures against competing terrestrial activities, the long lead time and large dollar amounts required by the former make some of the conventional short-term measures (return on assets or return on investment) inappropriate. Other methods of assessing and comparing such large-scale, long-term projects must be identified. For example, methods must be found to assess the value to a corporation as a whole of undertaking a space project in terms of its impact on the corporation's long-term, strategic position. In addition, the role to be played by government, through direct or indirect support, must be studied and discussed at some length as these projects begin to evolve.

Endemic to any space project is high cost. This constrains the scope of activities that can be undertaken due to the limited investment capability and risk appetites of financial institutions. Thus, some projects are best supported by the government, others by some joint arrangement between government and business, whereas others may enjoy the possibility of successfully competing for funds in the capital markets.

A SAMPLING OF COMMERCIAL SPACE PROJECTS

Until recently, the emphasis in space planning was on the "how" (e.g., how to send a television signal from an orbiting satellite to a certain place on Earth; how to build a rocket that is reusable). Today, the focus is on the "why" and that is primarily a financial issue. As discussed earlier, the process of industrializing space involves several key steps: (1) the establishment of a space infrastructure that allows one to enter into and operate freely in the space environment, of which the fundamental building blocks are the space shuttle fleet, the various space platforms, and the space station; (2) the development of a predictable institutional framework in which industry can operate; and (3) the creation of the financial structure required to support space-produced products and services.

The existence of the shuttle is already prompting a great variety of firms to investigate how the space environment may yield innovative products or, in some fashion, enhance a given product developed here on Earth. What follows is a brief description of some of the more intriguing proposals receiving serious attention.

Satellite Communications

Business is a major customer of communications satellite services. The greatest usage is in voice transmissions, but the greatest potential lies with data and facsimile transfer. (Satellite revenues from business satellite services are projected at $1.2 billion in 1985 and $2.8 billion in 1991.)

When describing the commercial impact from the communications satellite industry, one must take into account the earth stations that receive and transmit the satellite signals. More than three dozen companies manufacture earth stations, the leaders being NEC, Hughes Aircraft, Harris, Scientific-Atlanta, and IBM. Sales of earth stations were estimated at $130 million in 1983, with $900 million expected by the end of the decade.

COMSAT not only provides international communications services to U.S. common carriers through the satellites of Intelsat, it also leases Comstar satellites to AT&T. Seventeen U.S. communications satellites are in operation: AT&T, RCA, Western Union (four each); Hughes, Alascom (one each); Satellite Business Systems (SBS) (three). So far, the FCC has authorized 43 communications satellites to be launched between 1983 and 1986.

A very exciting, near-term commercial venture in communications satellites involves the use of very powerful satellite transponders that make it possible to send signals directly from satellite to customer (who will use small, private Earth stations). Satellite Television Corporation, a COMSAT subsidiary, will commence direct broadcast satellite service (DBS) in the United States next fall, using five satellite channels leased from SBS.[2] Broadcasts can be received with dish antennas only 2 to 2.5

[2]Satellite Television Corporation will orbit its own high-powered satellite in 1986.

feet in diameter, making broadcasts available to isolated areas. Among those awarded construction permits by the FCC, in addition to Satellite Television Corporation, are CBS, Inc., Western Union, RCA-Americom Communications, and U.S. Satellite Broadcasting.

The emergence of these technologies has produced entrepreneurs to challenge the established actors. For example, Orion Corporation, of Washington, D.C., is proposing to sell private satellite communications across the Atlantic Ocean. Unfortunately, this proposal runs counter to the Intelsat agreement that a member nation's communications satellites must not economically harm the organization's market share. Intelsat views Orion's proposal as a threat and is trying to dissuade the FCC from granting Orion an orbital slot. Meanwhile, Orion has obtained letter of interest from potential customers on both sides of the Atlantic and has already booked space on the shuttle for launch in 1986.

Other proposals under active consideration include transmitting communications satellite signals to very small, perhaps, Dick Tracy-style wristwatch radios; holographic teleconferencing; and the distribution of distress buttons that would signal a public-safety transceiver with the exact location of someone in need of assistance. A company called GeoStar has made a proposal to the FCC for a three-satellite system to be operational in 1987.

Space Processing

Beyond McDonnell-Douglas and Johnson & Johnson's experiments, other firms are seeking joint ventures with NASA to conduct a wide variety of experiments aboard the space shuttle. These experiments involve the production of semiconductor crystals, metals, glass, foams, and a myriad of other substances. Among those who have publicly discussed their plans are:

Microgravity Research Associates: Signed a joint agreement with NASA to produce semiconductors (specifically, gallium arsenide) in space. First scheduled test on the shuttle planned for 1986.

Westech Systems: Interested in using low-gravity conditions for the production of defect-free silicon crystals for semiconductors.

Boeing Aerospace: Considering construction of an in-space biochemical laboratory for the processing of greater quantities of high-purity biological materials than is possible on Earth.

Battelle Columbus Laboratories: Developing experiments to determine the effects of the absence of gravity on cell growth in animals and plants; researchers hope that this will lead to a better understanding of how to produce biological substances in space.

General Electric: Designing space experiments to see whether microgravity conditions will allow the fabrication of perfectly round latex spheres for placement in human capillaries for use in blood-flow experiments.

Ball Aerospace Systems: Negotiating with NASA to build a materials processing laboratory to fit into the shuttle's cargo bay.

Satellite Remote Sensing

American Science and Technology Corporation: Developing low-cost remote sensing satellites. Company has recently joined with a private firm, Space Services, Inc. (in a venture called Space America). Planning first satellite launch in 1985.

SPARX: Using a shuttle pallet, this venture is also planning to offer a commercial remote sensing alternative to Landsat.

St. Regis: Considering the use of remote sensing equipment aboard a space platform to monitor the health and size of the company's forest lands.

Transportation and Platforms

General Dynamics Convair Division: Aggressively seeking to commercialize their Atlas/Centaur ELVs in order to compete for communications satellites. The company is currently discussing arrangements with NASA for the use of government launch facilities and personnel, as well as the cost of purchasing parts and government-owned plans.

Space Services, Inc.: Its successful launch of a suborbital expendable launch vehicle, Conestoga, in 1982, marked the entry of the private sector into the launch business. The company is presently seeking financing for the construction of an ELV to launch light-weight, low-earth-orbit payloads. Plans call for Space America's remote sensing satellite to be the first payload.

Starstruck, Inc.: Developing its own low-cost commercial rocket, the company has already conducted over 30 engine tests and is planning a demonstration launch. They intend to offer a variety of launch services, from low-earth-orbit to geostationary capabilities.

Orbital Sciences Corporation: Currently has agreements with both NASA and Martin Marietta Corporation to develop, finance, and market a high-capacity upper stage for use in boosting payloads to higher orbits from the space shuttle.

Fairchild Industries: By 1986, it plans to orbit small, unmanned platforms that would remain permanently in space. Called Leasecraft, they would be rented to other companies for research and manufacturing. McDonnell-Douglas is negotiating with Fairchild to use Leasecraft for their drug-making process.

ORGANIZING FOR SPACE COMMERCE

Twenty-five years of space development, coupled with several successful space shuttle missions, has shown that space is a location which can accommodate people on an operational basis. With the prospect of a permanent manned facility, it is now possible to assure industry that this alien environment can be a suitable place for working.

The barriers to commercial space enterprise include onerous regulations and laws, enacted long before private investments in space were contemplated, which can discourage even the most enthusiastic entrepreneurs.

Added to these institutional problems are the risks and high costs inherent in space projects. Investments in research and manufacturing in space can be 10 to 100 times as large for equivalent ground-based facilities.

Yet, as demonstrated by the prospering communications satellite industry, the ultimate economic and social benefits from space enterprise can justify the initial risks. Certainly, as evidenced by the emergence of foreign competition, other nations perceive the potential of space industry.

In order to minimize the risks of doing business in space, there is an urgent need for the private sector and the government to forge a cooperative partnership. Among those steps that would greatly facilitate such cooperation would be assurances to entrepreneurs of long-range, consistent government policies; revision of laws and regulations that hinder space investment; expansion of NASA's cooperation with industry in basic and applied research; improvement in NASA's dissemination of information about space research to the business and university communities; and better definition of the responsibilities of government agencies in dealing with commercial space projects. Space, in our opinion, represents a new industrial revolution for the world.

REFERENCES

Allnutt, R. F., "The Commercial Era Dawns." *Aerospace* (published by The Aerospace Industries Association, Washington, D.C.) (Fall 1982).

Civilian Space Policy and Applications. U.S. Congress Office of Technology Assessment. Washington, D.C.: U.S. Government Printing Office, June 1982.

Encouraging Business Ventures in Space Technologies. Report by panel of the National Academy of Public Administration, Washington, D.C., prepared for NASA, May 1983.

Grey, J., *Enterprise.* New York: Morrow, 1979.

Gump, D., (ed). *Space Processing, Products and Profits 1983–1990.* Space Business News, 1983.

Haggerty, J. J. *Spinoff 1983*, NASA Office of External Relations. Washington, D.C.: U.S. Government Printing Office, 1983.

Kinnucan, P. "Space Business." *High Technology* 3 (10) (October 1983).

McDougall, W. A. "Technocracy and Statecraft in the Space Age." *The American Historical Review* 87 (4) (October 1982).

Naumann, A., and Alexander, G., (eds.). *Developing the Space Frontier*, American Astronautical Society, Vol. 52. San Diego: Univelt, 1983.

Policy and Legal Issues in the Commercialization of Space, prepared by the Congressional Research Service for the U.S. Senate Committee on Commerce, Science and Transportation, September 23, 1983.

"Space 25." *Spectrum* 20 (9) (September 1983). (Issue devoted to review of past space achievements and implications for future.)

17

America's Space-Based Missile Defense

James Wade

On March 23, 1983, President Reagan called for, as a long-term goal, putting an end to the threat of strategic nuclear ballistic missiles. The president recognized this would not be an easy task. There were many risks and uncertainties associated with achieving this goal. But he was also aware of our nation's finest resource — its creative and dedicated scientists and engineers. The challenge encompasses many technologies including those in space, and large-scale venturing by the private sector is a vital ingredient to our success.

What are the requirements and the opportunities that the president's strategic defense initiative holds? Much has changed in the past two and a half decades since the possibilities for ballistic missile defense were first considered. The threat has grown from the few hundred single-warhead missiles to thousands of multiple-warhead missiles. The situation is even more complicated with the proliferation of tactical and intermediate ballistic missiles. Nonetheless, with the impressive array of emerging technologies, a realistic defense is feasible, and many of these new capabilities are the direct result of far-sighted technology venturing from the private sector.

To place this problem in context, let me describe the flight of a ballistic missile. Such a flight has four distinct phases. In the first phase, first and second staging, the missile produces an intense and unique infrared signature. Next a postboost phase occurs during which the multiple vehicles (decoys) are deployed along the possible penetration edge. In the subsequent midcourse phase, warheads and penetration aids travel along ballistic trajectories toward the Earth's atmosphere. And finally, in the terminal phase, warheads and penetration aids enter the atmosphere and are reflected by drag.

The approach we are taking at this time is to engage the missiles in each phase of the attack, and to accomplish this we must have certain capabilities in each phase. We must have global full-time surveillance to perceive the attack moments after initialization so that we can engage and destroy as many missiles as possible during the slow lift-off stage, thus reducing the pressure during the later phases.

As the warheads are deployed and speed toward their target, we must be able to discriminate warheads from the decoys, lest our adversaries simply overwhelm our defenses with low-cost decoys. In order to eliminate the threat from incoming warheads, we must engage those ballistic missiles high enough in their terminal phase so that the intended ground attack on the United States is kept to a minimum. Finally, and most importantly, we must have an interconnected and survivable battle-management control system.

Now to accomplish all this, we have planned an intensive research and technology program for the remainder of this decade. Based on the success of this program, the effort may proceed to full-scale development in the next decade. I should emphasize that our strategic defense initiative is not a weapon development and deployment system, but rather a broad-based research effort to identify and develop key technologies necessary for effective strategic defense in the next decade. The research will be focused initially on technology for sensing and tracking missiles, technology for the weapons to be used against missiles and warheads, technology to support the control system, and technology to ensure the survivability and sustainability of the entire system.

What are some of the areas where investments and support are needed for the future? We have categorized our efforts into five broad technology areas. Each major area consists of technological development and a set of demonstrations and experiments.

The first area, and the one in which we plan the largest investments over the next five years, is designed to develop technologies for surveillance, acquisition, tracking, and damage assessment. We are particularly concerned about means and mechanisms for imaging objects in space. We must be able to discriminate warheads from decoys and debris. For the requisite acquisition and tracking we must have reliable radiation-hardened, large-format array infrared sensors in order to capitalize on the new technology. Although we are counting on new optical capabilities for detection, tracking, and discrimination, we must and will continue to pursue radar technologies as an alternative approach.

Direct energy offers a totally new capability to act nearly instantaneously over large distances. For this reason, this technology is generating high interest for boost phase intercepts. Although we have come far since lasers were invented some 20 years ago, we have far to go before we achieve the power levels and capabilities needed for ballistic missile defense. It is in the direct energy area that I most anticipate the major vectors for the future. We will be pursuing certain options for laser technology. The primary goal is to achieve high-power levels at much shorter wave lengths. We are interested in chemical lasers, such as oxygen and iodine, and electrically powered lasers such as free electrons and leximers. Particle beams for space applications offer a viable option, as they deposit their lethal energy in a manner making it very difficult for countermeasures.

As important as energy weapons themselves are, our building and manufacture of light space objects and sufficient monitors is just as important. Here I see a real opportunity to develop new industries. The United States currently has a limited production capability for space-qualified measures, and we are concerned about our ability to produce them. New approaches to this problem will be strongly supported for the next five years.

For later phases of the ballistic missile trajectory, we will consider kinetic energy weapons that destroy their targets by physically hitting them. The key here is to develop small nuclear missiles fit to kill warheads. And we will have to produce these weapons and delivery systems cheaply enough so that our opponent will not attempt to defeat our defenses by building more and more offensive missiles. Since we could proliferate our interceptors more cheaply than he can proliferate more offensive weapons systems, the loop is closed. Here we have interest in the new technology cannons, the high-velocity guns, and the so-called electric rail gun. If these devices could accelerate warheads fast enough, say in excess of 10 kilometers per second (21,000 miles per hour), they could provide an alternative to directed energy weapons for the boost-phase intercept. The challenge here is to develop small projectiles that can withstand the enormous acceleration (100,000 G's) during the launch phase.

Next, the critical technology needed for a ballistic missile defense program is survivable interconnected battle management and C^3 facilities (that is, command, control, and communications). But the hardware requirements are stressing. Our greatest need is for automated tools of battle management, primarily software development. Those who have been successful in developing video games have comparative advantage for solving these software problems. This is an area where the United States has a decisive advantage.

We must not forget the supportive systems and the technology needed for strategic defense. Before we proceed with our new weapons systems, we must fully understand our opponent's systems. We must develop the means to make sure that our own defense systems survive against the enemy's surprise attack. Finally we must consider our space logistic requirements. We must have the ability to place up to 100 metric tons (220,000 pounds) in a variety of orbits and to move such payloads from orbit to orbit as required. Furthermore, we must seek other ways to make available additional material for shielding and construction in space itself.

We recognize a critical need for innovation within a strategic defense initiative, and we have reserved possibly 5 percent of our budget for the next five years for looking at entirely new concepts. It is in this area that technology venturing can pay off. We are looking at the possibilities of using near-Earth resources and large amounts of materials from the ground. Consideration will be given, for example, for using raw materials for satellite shipping and construction requirements in space. The practical benefits both to our strategic defense objectives and to industrialization in space are vast and deep. As these programs develop over the next five to ten years, the private sector will provide the innovation to make such a new concept a reality.

To manage the space defense initiative, we are turning to very strong central management within the Department of Defense. Often there is a dichotomy between those who have the knowledge and those who have the authority. To avoid this problem, we are appointing a single strategic defense initiative manager who reports directly to the Secretary of Defense. This new manager will have central control of the budget planning and execution, including the ability to reprogram the resources from less promising to more promising technologies as we move down the road. This centralized control, with decentralized execution in government research organizations, will provide a powerful pivot point for this national initiative. I would like to place the president's strategic defense initiative in its proper perspective, as one of the most important technological programs the United States has ever embarked on. It is a great hope for the future that does not represent a deployment decision nor is it a substitute for current strategic countermeasures or for arms control. Rather, it will provide a technological demonstration affecting all deployment decisions in the future. I have every confidence that we can persevere together to make the president's goal a reality and give our children a safer world.

PART VI

LARGE-SCALE PROGRAMS: THE NEXT STEPS

18

Private/Public Venturing Activities and Opportunities

Emil M. Sunley

Joint ventures between private companies and government are not a new phenomenon. In the earliest days of our country, state and local governments teamed up with private industry to build a railroad and canal network that fundamentally changed the U.S. commercial system. Initially, these governments offered loans and grants to the railroad and canal companies and invested in their stock. The bankruptcies in the 1840s discredited many of these early joint ventures. In the 1850s, however, the federal government moved in and subsidized the railroad industry through significant land grants, loan guarantees, and sizable mail-carrying contracts. In fact, one of the federal vehicles for developing the railroad network, the Union Pacific Railroad Corporation, sets a precedent for the pursuit of policy goals through the sponsorship of a private or quasi-private corporation.

Indeed, there are a number of examples of such past joint public/private efforts. The Reconstruction Finance Corporation (RFC) was created by Congress in 1932 to stimulate industries out of the Depression. It played an integral role in the development of the aluminum, steel, and rubber industries during and after the war. The Tennessee Valley Authority is a well-known case of an attempt to create a privatelike entity to serve multiple federal policy goals: flood control, navigation, provision of hydroelectric power, and experimentation in regional planning. The Federal National Mortgage Association (Fannie Mae), a pioneer in the secondary mortgage market, is a federally chartered entity that was spun off into the private sector. Government participation in joint projects has ranged from granting direct loans, as with the RFC, to the awarding of tax credits, as with the recently enacted research and development (R&D) credit to stimulate private-sector research.

In more recent years, the federal government has stepped in at times of crises. The government bailouts of Lockheed, in 1971, and Chrysler, in 1981, though much criticized, were quite successful. In the case of Lockheed, the federal government guaranteed $250 million in bank loans. The Chrysler package was much larger, $1.5 billion of federal loan guarantees and $2.0 billion in forced financial concessions from employees, creditors, dealers, and others. The government made a tidy profit on the Chrysler deal from the sale of the warrants. However, the government has not always stepped in. Penn Central and Braniff Airlines were permitted to go bankrupt though the government later provided loan guarantees for Penn Central.

Three recent public/private venturing projects that are interesting in that they have affected or attempted to affect technological change:

1. *The Synthetic Fuels Corporations* was product of the Carter administration, created largely in reaction to the international oil market turbulence of the 1970s. The corporation is one of the federal government's vehicles for accelerating the development of alternative energy technologies and ultimately creating an entire new commercial industry.
2. *The Clinch River Breeder Reactor* project was initially authorized in 1970 and was killed in October 1983. It provides a good historical account of the development and commercialization of a technology. The project's failure should expose some potential pitfalls of government/industry associations.
3. *Comsat Corporation* was chartered in 1962. It is a unique example of public/private venturing. The private sector supplied 100 percent of the capital (there was no initial federal equity investment); the government supplied the technology and the monopoly on international communications. By most standards, Comsat is also the most successful of the three ventures.

From these examples I will draw some conclusions about the advantages and disadvantages encountered in government sponsorship of commercial venturing and about the forms of subsidy used. I will conclude with a discussion of alternative means of federal participation (in particular, tax expenditure-type incentives) and potential opportunities for future joint ventures.

THE SYNTHETIC FUELS CORPORATION

The policy goal of the federal government in establishing the Synthetic Fuels Corporation (SFC) was to direct energy producers toward nonconventional energy sources and to set the United States on a course of energy independence. The existence of a government role in the synthetic fuels industry in general, and of the SFC in particular, can be analyzed in terms of market externalities. The controlled domestic price of oil failed to fully reflect the proportion of the U.S. energy supply that was imported, the political and economic instability of the supplying region, and consequently the U.S. vulnerability to supply disruption. Following from this, it is argued, is the national security risk of another country, or group of countries, controlling such significant leverage on the United States and its economy.

Prior to the 1973 embargo, the private-sector synfuel industry was embryonic; although certain technologies existed, there was little commercial production. Even after the embargo, however, synfuels were still not economic mostly because technologies were untried and production costs uncertain. The industry cited such other impediments as poorer quality of the synfuel product, permit delays, and regulatory burdens.

On the other hand, it was hoped that government participation in the industry would accelerate a commercial capability that the government deemed essential to U.S. energy security. The SFC was thus created as a means of sharing the risk of developing that capability.

The corporation is entirely capitalized by the federal government and is run by a seven-member bipartisan board of directors. Directors are presidentially appointed and serve staggered seven-year terms. To allow it to function as much as possible like a private entity, the corporation was exempted from a number of federal agency regulations, and it is given the authority to pay employees more than other federal workers.

In it first phase, the SFC has available $20 billion to foster synfuel production. Using competitive bidding to select projects, the corporation is authorized to provide financial assistance via purchase agreements, price and loan guarantees, direct loans, and joint ventures (in order of decreasing preference). By June 30, 1984, SFC had to submit to Congress a comprehensive plan that reported on economic, environmental, and technological progress to date, as well as define the private sector and SFC roles in the future of the industry. Congress' acceptance of the plan triggers the second phase of SFC's existence and possibly an additional $68 million in support funds.

The corporation is not without its problems or its critics. The most notable characteristic of the SFC to date has been its inactivity (though this seems to be changing under increasing administration and congressional pressure). The only subsidy granted by the corporation is an $827,750 loan for a scheme to produce methane from peat bog in North Carolina. Compared to oil shale and coal, peat is just not a very sexy synfuel resource. Indeed there have been bills introduced in Congress to kill the corporation or to redirect its efforts.

The drop in the world price of oil has made synfuels even less economical. Critics contend that start-up subsidies can no longer be justified. In other words, the costs of fostering this infant industry have become too great, especially in the context of huge deficits. One of the original three synfuels projects, the Great Plains Coal Gasification Project, was the recipient of a $2 billion Department of Energy (DOE) loan guarantee (granted prior to SFC becoming operational). Its sponsors reported that the drop in oil prices yields a consequent drop in the price it can charge from $10 to $6, thus turning a $1.2 billion profit expected over its first ten years into a $773 million loss. Tenneco and American Natural Resources Company, the leaders in the Great Plains project, are currently appealing to the SFC for additional price guarantees.

The industry generally, then, is vulnerable to changes in the market price of substitute fuel. These changes can lead to an increase in the required government subsidy. But there is always some uncertainty as to the continuity of the government commitment. Can an industry make long-term financial investments based on policy goals that might come and go with administrations or with political tides?

In addition, the corporation is saddled with some potentially conflicting objectives that might hamper its work as a catalyst to synfuels commercialization. To achieve lofty commercial production goals mandated by Congress — 500,000 oil equivalent barrels per day by 1987 and 2 million barrels per day by 1992 — highly targeted efforts, using only the most developed technologies, are needed. At the same time, the corporation is mandated to pursue a diverse range of technologies, and to minimize environmental risks. Moreover, the corporation is to encourage "commercial" projects and not to subsidize research and demonstration projects. In any case, a change in emphasis of objectives across administrations might be sufficient to lessen industry's reliance on future subsidies.

It may well be that a more appropriate role for the Synfuels Corporation would be simply to guarantee the sale price of the synfuel and thus reduce a major risk factor in synfuel investments. Companies would present commercial projects to the corporation and the corporation would guarantee that the fuel would be purchased at the price of say $35 per equivalent barrel of oil. This would protect companies if the world price of oil should fall to $20 per barrel but would not protect them if the plant was profitable only at a price above $35 per barrel.

CLINCH RIVER BREEDER REACTOR

The Clinch River Breeder Reactor (CRBR) project was conceived in an era of optimism about the potential for nuclear energy: The government wanted to exploit commercial power production capabilities of nuclear technology and to realize a dream of "perpetual energy." Nuclear technology was born in the government domain to be used for national defense. But by the early 1950s, power replaced defense as its primary potential. By the late 1950s, light-water technologies were declassified and private utilities began producing power. The breeder reactor, a nuclear reactor that produces more in nuclear fuel than it consumes, held special promise: It would provide a limitless source of fuel for a nuclear-based energy network.

That the government would keep a role in the commercialization of the breeder seemed obvious. The federal government possessed the technology, still relatively untried, and it had already completed one experimental breeder, EBR-1. Of course, with nuclear power generally, environmental, safety, and national security consequences abound. The breeder reactor produces highly toxic plutonium, which, if uncontrolled, could cause epidemic cancers or be used in bombs. Finally, interest in the breeder was heightened, or perhaps exaggerated, by the New Frontier push for technological innovation. Other countries were already moving ahead with breeders, but it was important that the United States, as a nation, be *first*. Thus the federal government assumed the roles of both overseer and promoter.

The utility industry, though, was excited about breeder potential, especially those companies that already had light-water nuclear facilities and needed to be concerned about future fuel supply. Too, there was perhaps a political motivation for breeder enthusiasm among nuclear producers. If the Atomic Energy Commission (AEC) that regulated/promoted the industry was pushing for development of the reactor, producers would do well to support the cause.

The CRBR was authorized by Congress in 1970 at an estimated total cost of $500 million. The Liquid Metal Fast Breeder Reactor (LMFBR) would be pursued as a "demonstration program," and federal assistance was expected to be limited in nature and amount. The original statute authorized the AEC to provide $50 million, less the costs incurred in an initial research and development (R&D) phase; $10 million in fuel, free of charge; and up to $20 million in services and equipment. Congress contemplated financial risk born by private utilities and ownership by the contributing companies. But a four-party agreement signed in 1973 drastically altered the earlier plan. A consortium of utility companies would contribute a fixed amount of approximately one-half the estimated cost of the plant, $250 million. The federal government would assume all project cost overruns.

The consortium of utilities would be allowed royalty-free access to all patents and technologies gained in the process. They could assign personnel to the project to receive training. Through the board of the PMC, they also had a voice in managing the project.

By 1975, when the AEC went to Congress to request funds to meet the escalating cost overruns, opposition to the CRBR project began to stir. A number of circumstances had changed since the project was initially conceived. The price of electricity, which had been falling since 1950, started climbing, triggering a significant decrease in the rate of growth of consumption. Consequently, projected nuclear power requirements fell. In addition, uranium deposits expected to disappear by the 1980s, were projected to last much longer; the seemingly urgent need for breeder-produced fuel was abated.

Changes in the government bodies dealing with the nuclear industry and the Clinch River project also affected the project's future. The AEC, which had been a vociferous advocate of the breeder program, was reorganized in 1975 into separate regulatory and R&D agencies. The Joint Committee on Atomic Energy, which had a probreeder history, was dissolved under the 1974 congressional reforms. Jurisdiction of the breeder was spread to more than a dozen committees and subcommittees in Congress, opening the project to fire from all directions.

Finally, the cost overruns of the reactor appeared insurmountable. The perhaps "stripped down" $500 million skyrocketed to $1.7 billion by 1974, $2 billion in 1977, and finally to $4 billion in 1983.

In December 1982, Congress refused to appropriate funds to the Clinch River project for fiscal year 1983 until the DOE "vigorously" explored alternative means of funding that would increase private-sector participation and reduce federal budget requirements. A task force of utility industry and investment banking experts submitted to Congress a detailed refinancing proposal in August 1983. The plan contemplated private-sector investments of $1 billion of the estimated $2.4 billion necessary to complete the $4.0 billion CRBR.

The premise of the plan was that the Clinch River project would have served two distinct, yet inseparable, functions. On the one hand, it was essentially an R&D undertaking that could have been funded by the federal government. On the other hand, the reactor would have produced heat that could have been used to generate electricity that could have been sold to private consumers. In effect, the plan required that this production capacity be sold to private investors. However, the physical assets and the results of operations could not have been simply divided between those exclusively related to R&D and those exclusively related to production of electricity. Therefore, the plan outlined a joint venture requiring capital contributions from both the public and private sectors. Private investors would have contributed $1 billion approximating the cost of constructing a conventional coal-fired plant of comparable capacity. The government would have funded the rest, and each would have owned indivisible interests in the project in proportion to equity contributed.

The most significant and controversial aspect of the plan was the use of tax benefits to lure private-sector financing. Under the plan, the $1 billion private investment in CRBR would have been treated like any other private investment in plant and equipment. Private investors would have been organized as a partnership. The partnership would have received up-front ITCs (Investment Tax Credits) and depreciation deductions on equity contributions, but no income would have accrued to it for several years. The entire plan was supported by a list of federal guarantees — DOE contracts, principal and interest on the bonds, and the availability of tax credits and deductions — that perpetuated the government's assumption of all the project risk.

Financial experts generally agreed that the tax attributes of the plan would have been sufficient to attract private investors. However, adoption of the plan would, in effect, have removed the federal subsidy from the appropriation process; it would become instead an elusive tax expenditure. Though direct government spending would have been reduced by $1 billion, the deficit would not have been reduced by that amount because federal revenues would also have been reduced by the amount of special tax benefits made available to the private investors. The loss of budgetary control, coupled with the unchanged burden of risk, contributed to Congress' ultimate defeat of the DOE proposal and consequently the breeder reactor.

COMSAT CORPORATION

A look at the history of satellite technology exposes a strong case for government participation in its commercialization. In general, the communications industry has a tradition of private ownership subject to government regulation. With respect to international satellite communications in particular (as with the breeder reactor), most of the technology had been developed by the government. Bell Laboratories, RCA, and others had all engaged in preliminary conceptual satellite research. But not until the government developed launching capacity did the industry really commit to satellite R&D. In the early 1960s, then, private industry alone was not technologically capable of developing the global communication system contemplated by the government. The problem was if and how to transfer the capability to the private sector.

The government intended to achieve several national policy goals with the technology. By its very nature, the technology would have important foreign policy implications. The government envisioned a global communication network organized and led by the United States. As is often the case in these joint efforts, timing was crucial. Waiting for unaided commercial development would risk losing a leadership position. And in pursuit of "world peace and understanding," even the Third-World countries were to be included in the system despite the uneconomic cost of placing Earth stations in so many sites.

Debate on the deployment of the technology focused on whether control of the operation should rest in the public or the private sector, and if in the private sector, who should own it. Among private-sector alternatives Congress considered a separate, federally chartered corporation with several ownership options: widely held public stock, stock held exclusively by the existing international common carriers, or some combination of international carriers, domestic carriers, and/or equipment manufacturers. The compromise ownership plan adopted allocated 50 percent of the initial stock issue to the leading international carriers, whose expertise in the industry proponents thought would contribute to expedient development of the system. The other half of the stock was sold to the general public. Private capital to fund the corporation was amply available. In fact, the stock offered to the public was heavily oversubscribed. The board of directors of the corporation would have 15 members, 6 each elected by the two classes of stockholders, and 3 appointed by the president of the United States.

Today the rate-regulated segment of Comsat's business remains its largest, generating about 60 percent of its operating revenues. Given the authority from the FCC, the corporation has expanded into domestic satellite communications, other international services, and plans to enter the direct-broadcast satellite business shortly.

There are several issues pending before the FCC that would affect Comsat directly. Currently, the FCC allocates all international communications traffic between satellites (Comsat) and undersea cables (AT&T and other carriers). However, the commission would like to discontinue this practice. In addition, carriers have petitioned the FCC for direct access to INTELSAT satellites, eliminating the intermediary role of Comsat. Carriers have also requested new ownership arrangements for U.S. Earth stations in the INTELSAT system, a change that Comsat opposes. These regulatory changes clearly could have significant impact on the viability of Comsat. But whether Comsat itself remains a profitable entity under the pressures of increased competition is quite a separate question from whether it accomplished its statutory mandate of launching a commercial satellite communication capacity.

Several antitrust issues arise in connection with Comsat. Was monopoly, for example, the best/only means for initiating the system? Even if it was, does a monopoly on international telecommunications continue to make sense? Should the ownership of the new technology have been vested in the owner of the existing technology? How these issues may have furthered or stifled the development of satellite technology is difficult to analyze; it is even more obscured by the complexities of regulation. It is fair to conclude, nevertheless, that Comsat has proved to be an adequate means of achieving federal policy objectives through the private sector.

ALTERNATIVE MEANS OF FEDERAL PARTICIPATION

These three examples expose two primary caveats inherent in public/private joint ventures. First, as in unsubsidized ventures, unexpected changes in demand for a product, and consequently in the price, can adversely affect the profitability of a project. But in the case of subsidized/encouraged activities those unexpected changes in demand will increase the government subsidy required to make the process economic. This phenomenon played a role in the demise of CRBR. How the price drop in the international oil market has affected the SFC remains to be seen.

Second, changes in the cost to the federal government of a subsidy (as described earlier or changes due to other external costs), or changes in political climate, can threaten the continuity of government commitment to a project or industry. In addition the frequency of a subsidy's subjection to federal budget review affects the risk of relying on government support. Private industry cannot make accurate economic investment decisions based on changing government commitments.

The government can promote technology development in many ways. It can of course directly fund the R&D. This generates a problem of how best to transfer the technology to private interests and directly rests on federal patent policies and technology ownership. Both Comsat and the Clinch River project are examples of technology primarily developed by government being planted in the private sector.

The government can sponsor demonstration projects, providing full or partial funding, and coordinating private-sector participation. The EPA-sponsored "Godzilla" project, which developed and tested automatic garbage pickup technology, is an example of a successful demonstration project, and Clinch River, an example of a failure. Rand Corporation did a study of federal demonstration projects in 1976. It concluded that the most successful demonstration projects were those for which the technology was fairly well in hand, costs and risks were shared with private participants, an industrial system for commercialization already existed (for example, a manufacturing capability and a demand for similar goods), and there were no significant time constraints.

The government certainly fosters commercial technological progress through direct procurement. Adequate demand for a technological capability can make its pursuit economical. The entire weapons industry is, of course, an example of this. But Cray Research, Inc., a private-sector producer of large-scale, high-speed computer systems, might also provide an example. As for the defense industry, demand for Cray's product, at least in its first three years of sales, was created mostly by the federal government. National laboratories were the primary users of supercomputers. Unlike the defense industry, however, no guaranteed-purchase/cost-plus-type subsidies prompted the innovation. Now supercomputers are finding applications in industrial fields and are sold on the international market. Other producers are entering the market. The government's effective subsidy was its demand.

Finally, the government can grant industry or user incentives to encourage development and production of technologies. The price and loan guarantees of the Synthetic Fuels Corporation are designed to accelerate direct government involvement rather than other alternatives for joint ventures. Price guarantees, as discussed earlier, could remove the risk that a synfuel project would become uneconomical for the private investors because of a drop in world oil prices.

THE INCOME TAX: THE FEDERAL GOVERNMENT AS JOINT VENTURER

Up to now we have been examining what might be viewed as traditional joint ventures between private companies and the government: equity participation, loans, loan guarantees, grants, price guarantees, and government supplied technology. But in a very important sense the federal income tax makes the federal government a joint venturer in all investments. The government is, in effect, a 46 percent nonvoting shareholder; that is, the government shares in 46 percent of the profits and 46 percent of the losses.

Some qualifications obviously are needed. The government does not fully share in 46 percent of the profits because corporate taxable income, as a result of various tax incentives, is less than true economical income, though during periods of inflation corporate taxable income can exceed inflation-adjusted real income. Also, the government may not share in 46 percent of the losses because losses arising from an investment may not produce an immediate tax benefit if the corporation does not have taxable income from other sources. Under current law, a corporation with a net operating loss may carry that loss back 3 years and forward 15 to offset any taxable income in those years. If the loss is carried forward, the present value of the tax benefit is less than if the loss is usable currently. And a loss carryover may produce no tax benefit if it is not used during the carryover period.

Economists have long recognized that because the corporate income tax does not permit a full loss offset, the corporate tax discourages risk taking. The imperfect loss offset causes investors in risky ventures to require a very high expected rate of return. This may be particularly true of high-technology industries where investments have long lead times. These industries can expect to generate losses as each new generation of products is introduced.

By providing tax incentives, the government is in many ways a joint venturer. These tax incentives, often called tax expenditures, are substitutes for direct government expenditures. Tax incentives provided through the tax system range from the broad-based investment tax credit to many narrow incentives such as the jobs tax credit and the fast write-offs for rehabilitated historic structures.

In general, there are no inherent differences between direct expenditures and tax expenditures. Tax expenditures tend to go through different committees in Congress and are not subject to the annual appropriations process. This may provide greater certainty of continuing government commitment. Nevertheless, Congress frequently has reviewed and altered certain tax expenditures such as the investment tax credit. I am certain we can all identify direct expenditure programs that should be reviewed by Congress as frequently as the investment tax credit has been.

To insure adequate congressional review, Congress in recent years has tended to sunset new tax incentives. For example, the incremental tax credit for R&D enacted in 1981 expires at the end of 1985 unless extended by Congress. Moreover, certain tax incentives have controls similar to direct expenditure programs. For example, the Labor Department certifies workers as being qualified employees for the jobs tax credit. The Interior Department and state agencies must approve historic rehabilitations. Finally, tax expenditures generally are like entitlement programs: They are open ended and not subject to an appropriation ceiling.

Clearly one disadvantage of using the tax system to provide incentives is that these incentives may be of little value to companies that currently are not taxable. This would be too serious if all companies that currently are not taxable were "losers" that should go out of business. However, many companies that currently are not taxable are quite viable. They are start-up companies, rapidly growing companies with excess depreciation deductions and investment tax credits, or companies that temporarily have fallen on hard times, possibly because of factors beyond their control such as sudden shifts in world energy prices.

This has led Congress to consider various mechanisms for making tax incentives more evenhanded. For example, in 1981 Congress permitted companies that currently are not taxable to see their tax benefits to other taxable companies by entering into "safe-harbor" leases. These leasing transactions had little economic substance other than the transfer of tax benefits. Safe-harbor leasing clearly involved a problem of public perception. It was repealed as part of the Tax Equity and Fiscal Responsibility Act of 1982.

It is quite likely that Congress will have to turn again to the issue of how to make tax incentives available to companies that currently are not taxable. Tax incentives could be made refundable, that is, if a taxpayer does not have sufficient income or taxability to currently use available tax incentives the government would send the taxpayer a check for the amount of unused benefits. This approach has substantial business opposition partly because it would have the appearance of backing "losers" but also because refundability might undermine public support for tax incentives and result in a backlash against existing incentives such as the investment tax credit.

The 25 percent R&D tax credit, enacted in 1981, illustrates some of the difficulties in using the tax system as a mechanism for delivering government subsidies. To increase the "bang for the buck" from the credit, Congress provided that the credit applies only to incremental expenditures, that is, expenditures in excess of average expenditures during the prior three years. Once the credit was made incremental, special rules were required to handle corporate acquisitions and dispositions. Also, the credit was limited to regular corporations. If this had not been done, the incremental feature of the credit could have been gotten around by individuals who had no prior expenditures for R&D forming tax shelter partnerships so as to maximize the allowable credit.

Only expenditures incurred "in carrying on any trade or business of the taxpayer" are eligible for the R&D credit. For new companies, the trade or business requirement causes R&D expenditures incurred in the start-up phase of the company to be ineligible for the credit because until a company actually begins to produce products it is not carrying on a business. This "in carrying on" requirement also raises substantial problems in the case of research joint venture partnerships, even when all the partners are regular corporations.

The credit is not refundable. Thus, it does not provide a benefit for a company without tax liability before the credit. Because the credit is incremental, the effective rate of the credit, at the margin, is not really 25 percent. Once one takes into account the fact that additional expenditures for R&D this year will increase the base for the credit in future years thus reducing the available credit in those years, the effective rate of the credit is closer to 5 percent.

Proposals have been made to enhance the attractivenes of R&D partnerships. The "in carrying on" a trade or business test could be eased to a more liberal "in connection with" a trade or business requirement. This change would aid both new businesses and joint ventures. The R&D credit could be extended to individual investors. To limit the potential for tax shelter abuse, the credit would only be allowed to offset the tax on any income from that partnership in the same or subsequent year. This is essentially the rule that applied in the case of the new jobs credit, which was also an incremental tax incentive. The major change, however, that would enhance the attractiveness of R&D joint ventures is not a tax change but an antitrust one. The administration supports legislation in this area.

19

Private/Public Partnerships

Ken Gordon

Technology venturing covers the unique U.S. process through which innovation ideas and technology are carried through to products and services of economic value. The key in determining success in this venturing is competitiveness, particularly in the global economic arena. Whatever business you are in, you must be constantly alert to penetration of your markets by foreign-based competitors and to export opportunities for your products and services. Most of what we are talking about in this volume are ways to increase the effectiveness of that competitive process in our contemporary setting.

You may not believe it at times, but most of the governmental policy activities in Washington are aimed at increasing the effectiveness of that competitive process. I firmly believe that for our free enterprise system to operate successfully these days, there must be effective partnerships among government, business, and other institutions. Our task is how to do this.

In this paper, I will address the challenge we face reducing antitrust barriers to joint research and development (R&D), using R&D limited partnerships, and finally, some recent initiatives of the president's Commission on Industrial Competitiveness.

It should be no surprise to anyone in business that the United States as a whole continues to face significant competitive challenge in the world marketplace. The driving forces can be summarized in the following three categories.

First, there is the challenge of new technology. Scientific innovation is bringing about a restructuring of our economy. We are witnessing an explosion of new technology in such areas as information transfer, new materials, factory and office automation, and biotechnology, to mention just a few. These developments make many of our capital investments obsolete prematurely, as well as make our technical and managerial skills obsolete more rapidly. As mentioned in other chapters, these new technologies will call into question such existing public policies as antitrust laws and regulatory structures, as has already occurred in the communications sector and the shrinking basic industries.

The second challenge comes from new strategies of foreign competitors that successfully target and capture market shares in technologically intensive markets. Such targeting, pioneered by Japan, is characterized by combined government/private actions that usually include early identification of emerging technologies or markets; intensive R&D; protection of domestic markets; and attention to ensuring very efficient, high-quality manufacturing.

The third challenge stems from the changing demography of the United States combined with changing labor force demands. We are beginning to undergo a significant slowdown in the younger age groups and a substantial growth of the elderly population. Because of ongoing industrial and technical transitions, there are shortages of some critical skills and upcoming major needs for retraining. This presents a significant challenge to our educational systems that we have not yet met.

I am not going to present the data on the relative decline of the U.S. industrial position — we've all been saturated with such over the last two years. I will only cite a comprehensive study of the decline of the U.S. manufacturing industry released recently. This independent study was sponsored by eight major U.S. corporations. It was conducted by Data Resources, Inc., and directed by Otto Eckstein, the eminent economist. It documents the importance of the manufacturing sector in the U.S. economy, as well as its continuing decline. Relevant to the focus of this volume is that two of the more important causes of our current difficulties are judged to be the low investment in manufacturing and the high cost of capital for the U.S. relative to Japan. The U.S. cost of capital averaged three times as high as Japan over the last two decades. Unfortunately this leads to a requirement for much higher rates of return before investments in research or facility modernization can be justified.

This is not a political chapter, but I would be remiss if I didn't point out that the economy has improved considerably during the President Reagan's tenure. At present, interest rates are down, inflation has all but disappeared, and the economy is improving steadily. Federal expenditures have grown significantly for basic research. Unemployment dropped

3 percent last year and capital formation is growing again. However, we all know that there are clouds on the horizon because of large deficits and negative trade balances. The test of our leadership ability will be to overcome these difficulties without succumbing to the temptations of trade protectionism. Our use of protectionist measures to assist a specific troubled industry is more likely to lead to retaliation by other nations against a variety of different industries where U.S. exports are significant. I am optimistic about the United States. We have many opportunities open, and I continue to see many examples in both government and industry where needed changes have been recognized and actions started to move forward.

Extensive research and development has been unquestionably a critical contributor to the preeminent industrial position we have held for several decades. A strong technological base is still one area where the United States retains a competitive advantage, although we are often less effective in carrying that technology into commercial products, processes, and services. In the face of growing competitive threats, joint research is clearly one way to assure our future technical vitality.

The large-scale projects described in this volume often demand joint R&D. Two important facilitating mechanisms for joint R&D are undergoing changes in policy and/or application: antitrust barriers and financing through R&D limited partnerships. I refer particularly to joint R&D among private-sector firms, although there are increasing activities, both policy oriented and technical in nature, that involve joint R&D among companies, federal laboratories and universities.

Joint R&D among firms is not illegal under current U.S. laws. Because many types of R&D have become increasingly complex and expensive, cooperation has become an important avenue for lowering the cost and risk associated with R&D. The most visible example of joint research among firms in an industry is the Microelectronics and Computer Technology Corporation now headquartered in Austin, Texas.

Most corporations have a perception that antitrust laws are hostile to joint R&D efforts. Many in this volume mention antitrust as a problem area. Even when applicability is unclear, company lawyers will usually advise against taking the risk. Perceptions are aggravated by the fact that the defendant, if proven guilty, is automatically liable for three times the antitrust damage actually involved. The net result is that these perceptions may well deter the formation of some socially desirable joint R&D ventures that would contribute to the nation's competitiveness.

Support for reducing these antitrust barriers to joint industrial R&D has been building during the last three years in both government and industry. It is possible that legislative action may be taken this year or next to ease these antitrust constraints. A dozen or so bills have been introduced in Congress embodying several approaches. After joint development efforts by the Departments of Justice and Commerce, the Reagan administration submitted the National Productivity and Innovation Act of 1983. This bill addresses the adverse deterrent effects in two ways. First, the legislation would prohibit the courts from condemning joint ventures without first considering the potential benefits. This would reverse the burden of proof currently required. Second, the bill provides that participants in a joint R&D venture that has been fully disclosed to the Justice Department and the Federal Trade Commission would be liable for not more than the actual antitrust injury caused by the venture, not triple damages.

At the same time, the administration's approach does not adopt the requirement of some bills that the Department of Justice evaluate and certify in advance that a joint R&D venture does not include anticompetitive features. Such certification requirements would put the Justice Department in a regulatory role and probably lead to the establishment of arbitrary standards for market share and other measures plus, of course, create an administrative burden and more red tape. A second part of that bill would similarly modify antitrust restrictions on the licensing of the intellectual property that results from the joint research. The amendments say that licensing is not inherently inconsistent with antitrust laws and procompetitive benefits must be considered first. The threat of triple damages is also eliminated. The act would send a clear message that intellectual property enhances rather than impedes innovation and productivity and that antitrust enforcement must be appropriately sensitive to this fact.

With the current budget and political pressures, it is uncertain if an antitrust bill will pass this congressional session. But I understand some modifications are being negotiated at this time and this may increase the probability of passage soon. For example, a modified bill might include a "safe-harbor" provision that would clarify protection for joint R&D ventures properly disclosed in advance.

My own message to firms that could benefit from joint R&D is not to be deterred by antitrust concerns, as long as anticompetitive constraints are not part of the arrangements. The winds of change are blowing in terms of enforcement and rules clarification. Of course, that does not mean that joint R&D will be the savior for every industry facing foreign competitive pressures, but where it is useful it should not be avoided for antitrust fears. (The president's Commission on Industrial Competitiveness adopted, on February 3, 1984, a recommendation for legislative changes of this type easing such antitrust barriers.)

A further topic that strongly affects technology venturing is the ability to obtain the funds needed for conducting industrial research and developing products/processes. The part of the innovation process that translates ideas and technology into commercial products or services usually requires 90 percent of the cost, risk, and time of the complete innovation process. It is also the part that is most difficult to fund and benefits from the least federal incentives. The one method that has become widely used in the last few years for funding R&D is limited partnership. Bruce Merrifield has been an active proponent of R&D limited partnership for large-scale projects where internal funds cannot be made available. This mechanism is well suited for developments in the prototype stage about two to four years away from commercial operation. Investment funds are raised from limited partners and the general partner contracts to have the R&D carried out. The firm wanting the R&D done contracts in advance to buy the results, usually on a take or pay basis. Existing tax laws provide significant tax benefits which flow through to the limited partners. The unique advantages of the R&D Limited Partnership are that:

- It is equally available to declining as well as growth industries regardless of the internal cash flow position. It allows off-balance-sheet funding.
- It does not require the loss of equity ownership that results from venture capital funding.
- It can provide a scale of effort and a coalition of skills, facilities, and resources that are beyond the capability of any individual company alone.
- It can legally proceed into manufacturing if economies of scale in world markets are required; alternatively, it can license the technology for manufacture as desired.
- At commercialization it can continue as a research foundation; it can become a publicly traded stock; it can issue equity shares to its limited partners; or it can sell out and distribute cash.

Although the R&D Limited Partnership is not well suited to all venturing situations, I commend it to your consideration where some of these advantages apply. The Department of Commerce has published guidelines on establishing an R&D limited partnership, as have some of the major national accounting firms.

The last subject is the president's Commission on Industrial Competitiveness. President Reagan established this commission in the fall of 1983 to provide recommendations on public and private actions that will increase long-term competitiveness of U.S. industries both at home and abroad, with particular emphasis on high technology. I think many of the commission's concerns are relevant to the interests of this volume.

The commission consists of 30 people, mostly corporate chief executive officers (CEOs), but including several labor leaders, university professors, and one government official (Dr. Jay Keyworth, Science Advisor to the President). The commission chairman is John Young, president and CEO of the Hewlett-Packard Company. The commission has organized its investigations into the four areas: human resources, capital resources, international trade, and R&D and manufacturing.

Although its deliberations are in early stages, the first series of recommendations was officially approved by the commission at its third meeting. Most of these initial recommendations deal with increasing the incentives for industry to undertake research and development by reducing existing barriers, improving the potential economic returns, and ensuring the ability to benefit from the results of that research.

The first recommendation is to reduce antitrust barriers to industry joint R&D in the ways I discussed earlier.

The second recommendation concerns several tax incentives for R&D. In 1981 a 25 percent tax credit was enacted to stimulated increased R&D by industry, but it expires in 1985. The commission judged this tax credit a needed incentive and recommended it be made permanent. It also recommended investigation of the desirability of applying the credit to total R&D rather than just to incremental increases and recommended encompassing all R&D activities traditionally accepted by standard accounting principles. Also, current impediments for coverage of R&D by new firms and joint ventures should be removed. A separate proposed tax regulation (IRS Regulation 1.961-8) affecting R&D was judged on incentive to shift R&D overseas, and repeal was recommended. The commission also favored legislation that would create a preferential tax credit to encourage industrial investment in university research.

The third recommendation strengthens the protection of patents, copyrights, and trademarks as a way of ensuring that rewards to the originator of the research are received. Specific recommendations were identified to attack the problem of product counterfeiting, prevent the misuse of the Freedom of Information Act to obtain commercially sensitive industrial information previously submitted to the government, streamline patent laws and procedures in various ways, and extend the life of a patent to compensate for time lost in federal premarketing testing and review.

In the international trade area, the commission also recommended support for extension of the Export Administration Act in a way that protects our industrial sector, and encouraged an investigation of the feasibility of establishing a governmental data bank to provide market and related data to U.S. industry that would help assess foreign competitive markets.

The president's commission is just beginning to address some of the more difficult issues such as the retraining or relocation of displaced workers, mechanisms that can lower the cost of capital, and actions that might meet the special needs of the basic or traditional manufacturing industries undergoing a major readjustment process. The commission welcomes ideas and views on all issues. You can communicate directly. Share your ideas. Remember we are all in this competitive battle together.

20

The Roles of Private and Public Sectors in Large-Scale Public Works

D. Keith Dodson

This chapter centers on the history and future of large construction projects commercializing new or unproven technology. I do not attempt to segregate public and private projects and believe the concepts are generic to all large pioneer and/or technologically advanced projects. By their nature most projects of this type will utilize at least some public funding. Due to the high risk of unproven technology and size, and with the need for an alternative to Mideast petroleum as a part of our national defense, the majority of public projects for the remainder of this century will involve synthetic equivalent energy sources. The country needs these projects, and the only viable source of funding is the public sector.

The history of large new technology projects is not a success story. A consistent finding of researchers involved in evaluating large pioneering and/or technologically advanced projects is that severe underestimation of capital costs and time required for construction of such projects is the norm. One must question why this practice continues. Several explanations exist, not the least of which is that cost overruns are largely attributable to factors exogenous to a project. This includes such items as inflation, governmental regulation, and other unforeseeable events, such as strikes and bad weather. Those responsible have long used these as "scapegoats" if you will, to explain away reasons for large cost and schedule overruns.

Surprisingly enough a recent study conducted by the Rand Corporation for the Department of Energy involving data from 44 process-type plants considered "new technology" at the time of their construction, found that the combination of the above factors as a whole was only responsible for 27 percent of the misestimation error experienced on these projects. The study further found that even though these factors represented significant cost increases, estimators were adept at anticipating their magnitude. In other words, despite widespread belief to the contrary, cost increases as a result of inflation, governmental regulation, weather, and other items exogenous to a project could be accurately estimated within reasonable confidence levels established within the industry. I suspect an analysis of all large projects, in particular those involving "scale up" or pioneering of size, would reveal similar trends.

Even more revealing, the same Rand study concluded that the major factors that account for cost overruns can largely be identified *early* in the development of the technology, long before major expenditures have been made for design, much less construction.

The Rand study concluded that *process characteristics* and the amount of *definition* of the characteristics heavily influence the accuracy of estimates. This does not mean that a technology's being unproven is the direct cause of underestimation. Rather, it is the unforeseen design, engineering, construction, or start-up problems that unproven technology can run into, and that often require expensive redesign or repair, that are the real causes of misestimation. In other words, it is the implementation, not the process, that has caused these projects to run out of control. Through their hypothetical model, the Rand study deduced that the percentage of the total misestimated cost related to the accuracy of the estimate. This should come as no surprise. After all, the less that is known about a particular project, the more likely it will be misestimated.

Another component identified as a significant influence on cost estimating accuracy is the degree of project definition. As in the process characteristics, the amount of project definition available to the estimator determines to a large extent the accuracy of his estimate. It may help here to distinguish between process characteristics and project definition. *Process characteristics* deal with such items as feedstock characteristics, catalyst deactivation and impurity buildup, temperature tolerances, pressures, corrosive materials, abrasive materials, and solids/liquids/gas handling. *Project definition* deals with site location and layout, heat and materials balances, and equipment needs. For plants utilizing existing technology, these two items are less interdependent than that for new technology. This is to be expected; existing technologies have accurately proven process characteristics and only the details of location, equipment, and so forth need to be defined.

Additional explanation often cited for misestimation is optimism on the part of those responsible for pioneer projects and new technology.

The reasons for this are easily understood through the common example often cited in the literature. A large portion of research and development (R&D) is conducted by firms independent of the ultimate owner. In order to sell their product, R&D firms attempt to make it as attractive as possible. Even in firms where "in-house" R&D is conducted, it is argued that to justify and extend their existence, these departments "package" their technology in an as attractive as possible package. This often means overoptimistic estimates. Although widely held, there is no concrete evidence that this theory covers the large discrepancy between estimated and final costs. Perhaps if the initial estimate developed in the early stages of R&D were used as a basis for measurement, some merit could be accredited to this argument. However, I suggest, and the Rand study mentioned it in their findings, that the major concerns involved in the construction of plants involving new technology discount these types of estimates as being highly optimistic. Generally it is the unusual project where inexperienced personnel are involved that gets into this kind of difficulty.

Another excuse becoming progressively more prevalent today is poor management. It is important to distinguish here between two different classes of management. The first type of management is involved in the decision-making process of evaluating new technologies and their feasibility. They are involved in the decision as to whether a project proceeds from R&D to project design to construction.

The second type of management is involved in the design and construction of the project. (It should also be noted that in some instances the same personnel are involved in both management categories. The purpose here, though, is not to identify personnel but merely to separate functions.)

Now that we have separated the functions we can allocate the blame commonly placed on management. The first type of management is cited as not learning from past mistakes where estimates have been in error. They are accused of brushing over similar problem projects, citing that the situations involved are unique and not likely to occur again. Generally, they are strong in their belief that the project is viable because they developed the numbers. It is our experience that all projects have similar patterns within segments of the effort.

Close scrutiny of these patterns can discern undesirable trends at an early stage, but it is always difficult to get the owner to recognize that his project is in trouble. A portion of this excuse can be explained when one considers that most firms do not venture into projects that involve new technology very often. Traditionally most owners construct the majority of their projects using proven technology and on a relatively small scale. Thus, the excuse that the project is unique has some merit.

The second type of management, on the other hand, has quite a different share of the blame for cost overrun in most instances. One must question whether the second type of management is bad because the cost estimate is bad or whether the cost estimate is bad because management is bad. It is the old "chicken and egg" question and is compounded by recent findings that suggest it is not necessarily management that is responsible for performance but rather established company practices in place long before the current management was around. (Figure 20.1 summarizes the Rand study variables.)

It is obvious that serious underestimation of large pioneering and new technology projects is a major problem that must be dealt with if we are going to evaluate and plan new projects effectively. We have identified several causes for cost growth. These include incomplete or inaccurate process characteristics and project definition, overoptimism on the part of R&D efforts, and poor management.

The Rand study cited previously examined several of the factors associated with process characteristics and project definition. They intended that the results of their research would supplement current cost-estimating techniques rather than replacing them. Their findings strongly suggest that cost growth stems directly from low levels of information both about technical processes and about the project itself on an actual site.

A regression analysis was performed in an attempt to forecast final costs at early stages in the project's life. This regression identified five variables that influence cost growth for these types of projects (Figure 20.1). These five variables are the percentage of the project that represents new technology, the amount of impurities in the process, the complexity of the plant, the inclusiveness of the estimate, and the extent of project definition. It is interesting to note that the amount of impurities in the process statistically influences the accuracy of the estimate. The Rand study identified and measured several technical aspects of new technology plants in an effort to evaluate their relationship to estimate accuracy. They suggest that first-of-a-kind plants exhibit particularly high levels of design difficulties with impurity buildup and corrosiveness. Perhaps this is duc in large part to the lack of proper identification in the early stages of a project's development to correctly identify the proper materials of construction to support the process involved. Unacceptable

FIGURE 20.1
RAND STUDY VARIABLES

- % of New Technology
- Impurities in the Process
- Complexity of the Plant
- Inclusiveness of the Estimate
- Extent of Project Definition

FIGURE 20.2
TYPICAL PROJECT COST INFLUENCE FACTORS

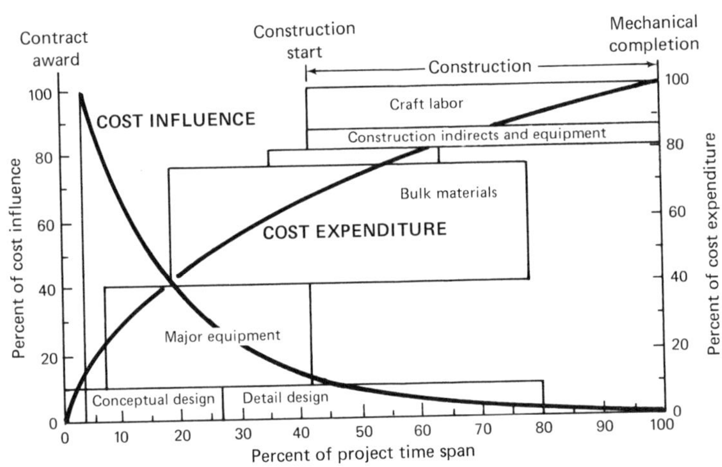

"stock materials" are used in the development of the estimate that later have to be replaced by more expensive corrosion resistant materials.

The result of the Rand regression is an equation that predicts final cost as a function of the estimate and the variables described earlier. The model developed statistically explains over 83 percent of the estimate error for the projects studied. And further, all the variables represent factors that are measurable early in the life of a project with some degree of accuracy.

These results are significant and should be evaluated carefully. I would suggest rather than interpreting these results as an end in themselves, they should be considered more as a stepping stone in the right direction. How comfortable should we feel using an equation early in a project's life to predict final costs based on values assigned to variables? Experience teaches that those values could be highly subjective. Also what is to prevent firms from adopting this model to their particular needs and ascribing biased values to the five variables mentioned earlier in a manner inconsistent with the study? Doesn't it get back to the same argument we addressed about optimism of research and development?

It is consistent with the Rand study (although not exactly parallel to their recommendations) that we must concentrate on more precise definitions for the variables just identified early in the stages of a project's life, so that accurate estimates can be based on better information. The use of the Rand model in preliminary stages of R&D is still a good method of assessing a particular technology's validity. But a regression model is not the whole anwser for improving estimating and subsequent cost of growth of pioneering and new technology projects.

Misestimation and cost growth can be minimized for new technology projects if the right action is taken early in the life cycle of the project to identify those elements (as discussed earlier) that influence the ultimate cost of a project. As shown in Figure 20.2, cost influence is lost early in the life of a project. The ability to control expenditures within reason is lost early in the design before any construction has actually started. The remaining 15 percent is represented by the pricing and productivity of both design and construction labor and the pricing of equipment and commodities. Importantly, the amounts or quantities of these items are predetermined by the previous three elements. This makes critical the necessity for "early-on" definition of process characteristics and project definition; otherwise the ability to control costs is lost. Generally, this process must take place in what we call the "rose-colored" portion of the project, where all are taken by the enthusiasm and where everyone is elated about the project going ahead and not concerned enough about where the project is headed. Generally by the time interest is stimulated all realistic influence over cost is lost. The time for realism is up front.

Some may argue that the nature of new technology makes it inherently difficult to define these characteristics. I do not dispute this argument but stress that it is imperative that we work to accurately determine and control a project's cost. After all, what man builds a house if he had not previously considered the cost? We must and can identify these elements.

What are the primary cost items of major process plants? Based on over 14 years of project cost experience, the point I want to emphasize is that commodity materials, that is, nonequipment items, represent on average the largest expenditure. This element is generally of less concern during the developmental period of the project, which, as we have depicted, influences cost greatly. Also, as the Rand study pointed out, many times the processes are more corrosive or difficult to contain, thus requiring more exotic material that directly affects these transfer-type materials more than the rest of the plant. An early emphasis on this element is a major point of cost control for the entire project. Many tools exist for early control of corrosion or containment, and it is incumbent upon us to use them to improve efficiency in such large-scale projects.

Although the Rand study does not identify government regulations as a major contributor to cost overruns, I believe that government agencies' *lack of clear direction* has contributed heavily to costly delays and additional capital expenditure in many projects in which I have been involved. We need to make good rules and stick by them. Change is inevitable, but we need to manage it. Let's not decide that all bathrooms have to go on the second story after we are under construction on the house and have poured the foundation. Nuclear power plants and their almost unbelievable increase in cost and time of construction are good examples of the impact of this philosophy. (Government and industry have generally avoided such catastrophic tail chasing on other projects.)

Large pioneering and/or new technology projects can be built economically in this country. Industry, government, and labor all must make concessions and sustain a complementary effort. We have identified the problems, and now must make a concerned effort to manage them.

The larger the project the more difficult the task, but risk reduction will lead to more projects and increased opportunities for Americans.

21

Joint Technology Venturing: New Institutional Arrangements, The Next Steps

Frank P. Davidson

If "technology venturing" is to be optimized as a central national activity, we shall have to look at it as a multifaceted, multilevel task. Full scope must be given to small, individualized enterprises. But we must also take into account the requirements of extensive engineering systems that can span a continent or, in some cases, the planet. Nor should we neglect the challenge to the moon, the asteroid belt, and outer space.

This volume is important. For here, convened together in one place and, I daresay, in one spirit, are leading representatives of the "three estates" of our realm: business, government, and academe. We do not indulge in the heady rhetoric, unfortunately so endemic to traditional political discourse, of the conflicting interests of the public and private sectors. The only pragmatic issue is how all the sectors of national life can best contribute to the shared purposes of the republic.

Beyond the belief in individual and national independence, we share a determination to protect the public health, advance the national prosperity, and do everything necessary to secure the safety of the republic and of its allies. The American Society for Macroengineering can do much to assist us in attaining these objectives. This new interprofessional organization will do much to explain, both to the professions and to the public at large, the immense opportunities that American engineering knowledge can offer, once the issues and impacts are evaluated and understood.

It is not a question of futuristics. The challenge we face is one of present judgment and choice. As the advanced technologies of computer science, telecommunications, ballistics, and bioengineering are successively applied to industries that have been classified as "low tech," we shall see a surprising neoindustrial economy sprout wings and take off. Individuals will differ on priorities and estimates, but few will quarrel with the proposition that vast and dramatic improvements can be made in such infrastructural amenities as water supply, transportation, and the development of the resources of the oceans and outer space.

As we grapple with specific options, it may soon dawn on us that recalcitrant problems are not caused by lack of technical or scientific knowledge, but by the hard fact that our institutions and attitudes were shaped in a more parochial age and that we must improve our teamwork in decision making and implementation if we are to be more successful in upscaling "technology venturing."

American managerial and legal talent assisted the nations of Europe, after the disasters of the Second World War, in building the many confederal institutions of the Common Market. Meanwhile, our consultative procedures with respect to North American development remained on a nineteenth-century basis. It is not time to examine the possible advantages and impacts of a common understanding with Mexico and Canada on such issues as energy, water supply, and transportation? Texas boasts a leading world center for the study of water resources: The International Center for Arid and Semi-Arid Land Studies (ICASALS), founded in Lubbock on the initiative of that farseeing geologist, Dr. Grover Murray. The University of Arizona and other famous institutions in the Southwest also make important contributions to the generation and analysis of policy options on such matters of national and international concern. Or must all initiatives come from Washington, D.C.?

One of the odd facts of our political system is that we place almost superhuman burdens on those elected to high office. I remember being surprised by the remark of a member of my own family, who headed the water supply department of a major city in the East, that he barely had time to sign all the contracts placed on his desk and that there was no way for him to read in detail the commitments that he was obliged to authorize: Delegation was a physical necessity.

If there are indeed to be new directions in the national life, we must have incubators of thought and reflection far from the "pressure cooker" caldrons of bureaucratic frenzy. And if large-scale engineering-management, an American skill much more employed abroad, is to be harnessed at home so that our economy's performance may more nearly match its promise, then a "think tank" devoted to this issue is not only advisable, it is essential and long overdue.

In this perspective, we can all salute the authorities at the University of Texas who have had the perspicacity and the pluck to design a Center for Macroprojects, to be located at Austin and to be a focal point for the efforts of several institutions in the Southwest concerned with this urgent national issue. Nor am I unmindful of the good sense of the administration in selecting, as the center's first associate director, a recent doctoral graduate of the Massachusetts Institute of Technology, Steward Nozette, an aeronautical engineer who has won his spurs at the University of California's California Space Institute. Dr. Nozette is too young to know all the excellent reasons for avoiding hard problems and embracing inaction as a way of life; in this proclivity, I think he represents the youthful spirit, if not the chronology, of my cocontributors to this volume. He and his able peers have my confident good wishes as they embark on a task of vital interest.

What must happen, if this nation is to rejuvenate itself and regain the posture of leadership and confidence to which its citizens aspire, is a shift in our modes of discourse and organization. We cluster around identifiable tasks, however extensive, forming "across-the-board" teams made up of persons with the resources needed for specific performance.

In the very difficult days of World War II, Lord Louis Mountbatten coined the phrase "combined operations" to characterize the necessary, but hitherto unpracticed, teamwork among all arms of the allied forces. My own regiment, the Fort Garry Horse, suffered its greatest casualties of the campaign because of inadvertent bombardment by "friendly" air forces. Perhaps we now need a civilian equivalent to wartime "combined operations." This is not to say that we need *less* competition; but we do need *more* cooperation. If our free enterprise system is to survive and prosper, we must as a nation change some of our habits and even some of our hallowed rules of the game. On matters vital to the Republic, groupings of business people, civil servants, and possibly uncivil *savants* can meet for lunch and plan those grand enterprises to which our capabilities, and our requirements, entitle us.

What is the "menu" that I am proposing? My own biases have been set forth in a recent volume entitled *MACRO*, and no doubt many experienced people will feel that I have exaggerated, in some respects, the ability of our political, legal, and social systems to coalesce sufficiently to make the sacrifices entailed by what the French have always called felicitously, "les grands travaux." But look at what we have already accomplished, and with more slender means. Did not Governor DeWitt Clinton serve the long-range welfare of the Republic when he accelerated the opening of the West by building, in six short years (1819–25), the Erie Canal? Was it not an American entrepreneur who, in 1867, succeeded in laying a transatlantic cable? Is it not time for the nation that first sent astronauts to the moon to send new ones back not just to pick up some rocks but to build the hotel?

Much as I enjoy and appreciate academic meetings — we can recognize something of ourselves in Arthur Koestler's great parody in his novel, *The Call Girls* — should we not recall the Roman reaction to water shortages when their legions conquered the province of Africa? The army, having accomplished its first task, was set to build long-distance aqueducts, which, in turn, would provide the basis for a flourishing agricultural civilization. Not since the Roman days has Africa been able to administer such a resounding defeat to the encroaching Sahara Desert. Should we not, in our modern Republic, look with a Roman eye on the impending water shortages on this continent? Or must we be content with the United Nations' reaction to the starvation and drought in the Sahel, when a conference and publication schedule on desertification had to serve as an alternate to decision and construction?

The ancient Greeks were fascinated by scientific speculation but lagged, on the whole, in the application of scientific knowledge to daily tasks. Is there not a lesson in all this for our own policies, where each governor, in imitation of Washington practice, appoints a Council of Science advisors but often neglects to form a countervailing — but synergistic — Council of Engineering Advisors? I remember sitting in the office of the late, greatly missed George Rufus Brown when he assisted President Lyndon Johnson in formulating the great decision to conduct a public study of alternate routes for a second trans-isthmian canal in Central America. Today, we not only have new and more economical methods for building canals, but there is the opportunity to move even very large ships by mechanical means, along specially contrived roadways. I trust that the nascent Center for Macroprojects will not only collate information on such possibilities but will proceed to the formation of intersectoral teams for evaluation and the building and testing of models.

As the opportunities before us become part of the public consciousness, it will appear outmoded to divide industrial policy between the aerospace advocates of California and the heavy construction industrialists of the Midwest and South. The advanced aeronautical knowledge of California firms must now be applied to improving the terrestrial and ultimately the intercontinental transport of people and goods. Although Dr. Robert Slater's advanced concept of hypersonic coast-to-coast subway may present difficult problems of finance and authorization, Americans will not wish forever to be "catching up" with achievements in Japan and Europe. A "great leap forward" may be within our grasp, here at home, if we can but remember our history and evaluate the reach of present capabilities. Nor should the thrust into space be regarded as the private preserve of one section of the country. If we are indeed to use the moon as a base for the mining and fabrication of materials needed for the coming systems of space technology, our heavy construction industry will be needed to supply knowledge — and knowledgeable people — for building protected havens and serviceable transport and energy links on and below the lunar surface. These are all national tasks and, in appropriate cases, we shall wish to associate our neighbors on this continent, and our far-flung allies and trading partners, in the great and pragmatic adventure as humankind learns to cope with and benefit from the environment of outer space.

I take it that the Center for Macroprojects will serve as a focal point for a new and benign coalition of interests so that decision making on such matters may be reality oriented and so that our youth can be guided and trained in the multidisciplinary technologies and tasks that will build tomorrow's world.

Recently, I had the honor of visiting the vast base established at Prudhoe Bay, hundreds of miles north of the Arctic Circle, for the extraction of petroleum products and for their transport, via the Alaskan Pipeline, to the energy-hungry markets of the south. The policy implications of this achievement for our continental energy supply have been ably discussed as early as 1979 in a paper entitled "Markets for Alaskan Oil" by Professor Richard Mancke and his colleagues, published by IPC Business Press: There is an inescapable linkage between the arrangements for importing oil from Mexico and the authorization for oil exports from Alaska to its natural markets in the Far East. I am not expert in these arcane affairs, but no doubt the staff being assembled for the new center at Austin will grapple with such conundrums and suggest viable answers that can strengthen both our prosperity and our allies.

It is time to disaggregate our thoughts, just as we must begin to disaggregate such misleading statistics as "gross national product." The latter calculations include the income of our funeral homes, hospitals, and charitable foundations — all very worthy and necessary enterprises, but not in all cases reliable indexes of the national health! Recent studies have suggested that to build a home nowadays, more than 30 percent of the cost is taken up by paperwork of all kinds. Must we not increase that portion of the national effort devoted to construction — and to construction that will be of enduring value? Is it not high time for "technology venturing" on the grand scale, for a national commitment to excellence in macroengineering, for the preparation of *cadres* of young people trained in the indispensable skills of engineer-management?

If Peter Vajk's challenging phrase, "Doomsday Has Been Canceled," is to stand the test of time, the future itself must be leveraged to channel energies toward the selective implementation of the best programs that our current engineering capabilities can present. America has never been comfortable being "second best." I am sure that our Macrostate will lead the way to the greater future and the larger vision that has always been the hallmark of our heritage and destiny. *Ars Celere Artem.*

APPENDIXES

A
Technology Venturing Highlights

Eugene B. Konecci, George Kozmetsky, and Raymond Smilor

A NATIONAL AND INTERNATIONAL FOCUS

1. Over 150 programs or initiatives have been developed by the states for high-technology and economic development. The states have taken a leadership role in this area in the United States.
2. The major source for R&D funds is still the federal government. Those states receiving the largest federal obligations for R&D have taken the lead to initiate high-technology development. The major exceptions are North Carolina, Georgia, Minnesota, and Utah.
3. State financial support for technology development is difficult to identify and tabulate.
4. Fifteen states have passed high-technology education programs.
5. According to a *Venture* magazine survey, there are 50 high-technology highways in the United States. Three are mature high-technology centers:

    Silicon Valley in California
    Route 128 in Massachusetts
    Research Triangle in North Carolina

    Twenty-one states are developing 32 high-technology centers as follows:

California	3
Maryland	2
Florida	2
New York	1
Texas	3
Virginia	1
Ohio	2
Pennsylvania	2
Washington	1
New Jersey	1
Colorado	2
Illinois	1
Alabama	1
Arizona	2
Louisiana	1
Minnesota	1

Utah	1
Georgia	1
Indiana	1
Oregon	2
New Hampshire	1

Fourteen states are developing 15 emerging high-technology centers as follows:

Florida	1
New York	1
New Mexico	1
Ohio	2
Tennessee	1
Michigan	1
Rhode Island	1
Wisconsin	1
Oregon	1
Oklahoma	1
South Carolina	1
Arkansas	1
Maine	1
Vermont	1

6. A majority of university-corporation technology programs are with private universities. Public university-corporate programs are emerging.
7. The capital venture industry consists of the following six sectors:

   a. The private/independent venture capital limited partnership.
   b. Business development companies.
   c. Corporate venture capital subsidiaries.
   d. Small-business investment companies.
   e. R&D limited partnerships.
   f. Specialty funds.

   1983 was a banner year for capital venturing; over $3 billion of capital was raised — more than double that invested in 1982.
8. Supercomputers have become an active area of international competition for scientific preeminence. Japan, with its two supercomputer programs, is in competition with the U.S. Defense Advanced Research Projects Agency (DARPA) in the Department of Defense, the British collaborative research program, a European consortium, and a possible European Strategic Program for Research and Information Technologies.
   1983 was the first year that the Japanese produced and delivered an operational supercomputer (vector machine). They are planning world distribution during 1984.

9. Newer joint ventures are being initiated by U.S. corporations for high-technology R&D including

    Microelectronics and Computer Technology Corporation
    Semiconductor Research Corporation
    Stanford University Center for Integrated Systems
    Microelectronics Center of North Carolina

   Each has its own unique constituency, mode of operation and organization, and distribution of technology.

10. A survey conducted by the National League of Cities and the U.S. Conference of Mayors has found that the following infrastructure facilities are in generally poor condition:

    a. Streets and roads.
    b. Sidewalks and curbs.
    c. Stormwater collection and drainage.
    d. Sewage collection.
    e. Bridges, overpasses and viaducts.
    f. Wastewater treatment.
    g. Water distribution.

11. The congressional Budget Office has estimated that a capital investment of $53.4 billion of 1982 dollars annually is required through 1990 for the development of new as well as the repair of old public infrastructure systems. The federal share is estimated to be $28.2 billion annually of which about one-half is for highways.
12. The Association of General Contractors has forecasted that over $3 trillion will be required for rebuilding America's infrastructure system, which includes roads and highways, bridges and dams, and sewage and water systems.
13. Federal outlays for aerospace products and services are estimated to be $41 billion in 1984. Of this amount, over $34 billion is for DOD and about $7 billion for NASA. DOD estimates include $27 billion for aircraft and over $7 billion for missiles. NASA spending estimates include $4.6 billion for research and development, $1.2 billion for research and program management, and $129 million for construction of facilities.
14. Sales in current dollars for major aerospace companies for 1982 were over $74 billion of which $32.5 billion were non-U.S. government sales. Sales for 1980 were $58.4 billion of which $31.8 billion were for non-U.S. government sales.
15. Most macroprojects (projects over $100 million) were launched during the 1970s throughout the developing world. They depended on transnational

partnerships and arrangements for capital, technology, management and market access. Internationally, there are over 3,000 companies to provide these services.

16. According to Kathleen J. Murphy (*Macroproject Development in the Third World*, Boulder, Colorado: Westview Press, 1983), there were over 1,615 planned macroprojects in the 1970s in the world. Cumulatively, these projects required over $1 trillion of capital. U.S. firms won the largest contracts, e.g., over $10 billion each.
17. Major worldwide changes are underway in the macroproject marketplace. During the 1970s, the marketplace was for metal extraction and processing. In the early 1980s, it was for oil refineries, oil and gas pipelines, and coal and synfuel projects. The emerging macroprojects are developing around worldwide infrastructures: crosscountry roads, water supply and sewer systems, and irrigation systems. Foreign exchange and internal economic conditions are requiring changes in transnational partnership arrangements. There is increased pressure for local partnerships and for more active participation of local companies in the host nations.
18. U.S. relative share of world GNP declined from 33.7 percent in 1960 to 30.2 percent in 1970 and 21.5 percent in 1980. Japan's share increased from 2.9 percent in 1960 to 6 percent in 1970 and 9 percent in 1980. The United States still has the largest single-nation relative share of world GNP.
19. The Japanese study "Japan in the Year 2000" estimates that the world GNP in the year 2000 will be 58 percent in the advanced economies (United States, Japan, EEC, and other OECD nations), 20 percent in developing economies, and 22 percent in the socialist economies (USSR, Eastern Europe, and the People's Republic of China).

A TEXAS FOCUS

1. Diversification of the Texas economy by expanding the industrial base through technology is considered by many to be of paramount importance during the next two decades.
2. Eighty percent of the current high-technology firms in Texas are located in the following SMSAs:

Dallas/Fort Worth	37.7 percent
Houston	28.4 percent
Austin	8.4 percent
San Antonio	5.5 percent

The decision by Microelectronics and Computer Technology Corporation (MCC) to locate in Austin has given an important boost to the Austin/San Antonio Corridor as well as to the entire state of Texas.

3. Each of the high-technology SMSAs has its own developing technological characteristics as follows:

   Dallas/Fort Worth
   - Communications equipment
   - Electronic components and accessories
   - Measuring and controlling instruments

   Houston
   - Measuring and controlling instruments
   - Electrical industrial apparatus

   Austin
   - Electronic components and accessories
   - Measuring and controlling instruments
   - Communications equipment
   - Computer software

   San Antonio
   - Surgical, medical, and dental instruments and supplies
   - Measuring and controlling instruments
   - Electric lighting and wiring equipment
   - Office automation

4. Since 1979, high-technology firms have sprung up in Bryan/College Station, El Paso, Waco, Killeen-Temple, Lubbock, and Wichita Falls SMSAs.
5. More than 20 percent (115) of the *Inc.* 500 ranking of fastest growing small companies in the United States are found in Texas. Texas' 39 firms in high technology are second to California's 84 companies listed in *Inc.*'s ranking.
6. The governor of Texas strongly supports the concept of economic development. Governor White's personal role in attracting MCC to Texas is widely acknowledged and studied by other states.
7. According to Dr. Harden Wiedemann, executive director of the governor's Office of Economic Development, "Texas is reputed to have the best small business climate in the United States. . . . Although its official support for small business isn't as great as that of other states, a productive labor force and a fairly low tax burden (are) included on Texas' list of positives."
8. Texas has a mature R&D base in its current industries and outstanding research facilities in its universities. In addition, it ranks 3–5 in the following areas: R&D funding received from the federal government, R&D conducted by the private sector, number of scientists and engineers, and total dollar amount of R&D funding for Texas universities. But, Texas was still 41 percent below the national per capita average of public and private R&D — $216 versus $126.
9. One of the more comprehensive community action programs in Texas is the San Antonio/Austin Corridor program. Mayor Henry Cisneros is to be given much credit for its conception and development. It is a model of

intracommunity involvement and integration of policies. The entities involved are the city councils and mayors of San Antonio and Austin, their chambers of commerce, high-technology companies, private foundations, and private and public colleges and universities in both cities. Communication through selected meetings, television, and public press coverage has been outstanding. The program will do much to establish San Antonio as a "lightning rod" for biotechnology in Texas and the nation.

10. The workshops held at the University of Texas at Arlington and the University of Texas of the Permian Basin benefited from the example of the San Antonio/Austin Corridor Program. Currently, an Institute for Robotics and Artificial Intelligence is emerging under the aegis of the University of Texas at Arlington in the Fort Worth/Arlington/Dallas Corridor. The West Texas communities of Midland and Odessa are considering establishing a Center of Energy, Technology and Economic Diversification at the University of Texas of the Permian Basin. Other Texas community institutes and centers are currently being contemplated.
11. Capital venturing has a foothold in the state, but Texas has the need and ability to expand this base. As much as half of the capital venture funds raised in Texas may be invested in companies outside the state. In addition, much of the capital venture funds have been invested in oil and gas ventures, rather than in emerging technology businesses.
    According to Edward Vetter, venture capitalists "must bring more to the party than money." They need to bring "the experience that can be injected in the portfolio company . . . (including) hands-on operating experience in all the areas of a company. . . . The hands-on support can best be provided from the local level."

B

Summary of the State Programs for High-Technology and Economic Development Initiatives by Ranking of Federal Obligations for Research and Development as of 1983

Ranking by Federal Obligation for R&D	State	Percentage of Total Federal R&D	Total Number High-Technology Economic Development Initiatives	Number of Initiatives: High-Technology Development	Task Force	High-Technology Education	Labor/Technical Assistance	Capital Provision Assistance	General Industrial Development
Federal Obligation for R&D above 20%									
1	California	23.6	3	1	—*	1	1	—	—
Federal Obligation for R&D 5–19%									
2	Maryland	8.3	6	—	1	—	2	3	—
3	Massachusetts	7.2	5	1	—	1	1	1	1
Federal Obligation for R&D 3–4.9%									
4	Florida	4.8	7	3	—	1	—	—	3
5	New York	4.6	5	2	—	—	1	2	—
6	Texas	3.7	4	1	—	—	1	—	2
7	New Mexico	3.6	3	2	—	1	—	—	—
8	Virginia	3.4	3	1	1	—	—	—	1
9	Ohio	3.3	4	1	—	—	—	2	1
10	Pennsylvania	3.2	5	2	—	1	1	1	—
11	Washington	3.1	2	1	—	1	—	—	—
Federal Obligation for R&D 1–2.9%									
—	District of Columbia	2.8	—	—	—	—	—	—	—
12	Tennessee	2.6	3	2	—	1	—	—	—
13	Missouri	2.4	8	1	—	1	2	2	2

Appendix B (continued)

Ranking by Federal Obligation for R&D	State	Percentage of Total Federal R&D	Total Number High-Technology Economic Development Initiatives	Number of Initiatives					
				High-Technology Development	Task Force	High-Technology Education	Labor/Technical Assistance	Capital Provision Assistance	General Industrial Development
14	New Jersey	2.3	1	—	1	—	—	—	—
15	Colorado	1.9	4	1	1	—	1	—	1
16	Illinois	1.7	6	2	1	—	2	1	—
17	Alabama	1.7	1	—	—	—	—	—	1
18	Connecticut	1.4	6	3	1	1	1	1	1
19	Kansas	1.4	2	—	1	1	—	—	1
20	Arizona	1.1	4	—	—	1	1	1	1
21	Michigan	1.1	10	8	—	—	1	1	—
22	Louisiana	1.0	2	—	1	—	—	1	
Federal Obligation for R&D 0.6–0.9%									
23	Minnesota	0.9	2	—	—	1	—	—	1
24	Utah	0.9	1	—	—	—	—	—	1
25	Nevada	0.8	1	—	—	—	—	—	1
26	North Carolina	0.8	4	1	—	2	—	—	1
27	Georgia	0.6	3	1	—	2	—	—	—
Federal Obligation for R&D 0.0–0.5%	All other states	5.60	48	4	2	1	12	11	18
	Outlying areas	0.1	—	—	—	—	—	—	—
	Office abroad	0.1	—	—	—	—	—	—	—
	TOTALS	100	153	38	9	17	27	27	37

*Data not available.

Source: IC² Institute, University of Texas at Austin. Data from National Science Foundation (NSF) and Office of Technology Assessment (OTA), 1984.

C
America's 50 High-Technology Highways

State	Area	Participants: Universities, Government Entities, Base Companies	Government Agency	Mature High-Technology Centers	Developing High-Technology Centers	Emerging High-Technology Centers
California	Santa Clara County: "Silicon Valley"	Stanford, Fairchild Camera and Instruments, Hewlett-Packard, Apple Computer, Intel, National Semiconductor	California Department of Economics and Business Development, Sacramento	×		
	Orange County	University of California-Irvine, California State-Fullerton, Long Beach State University, North American Aviation, Ford Aeroneutronics, Baker International, Xerox, Cannon	Economic Development Corporation of Orange County, Irvine		×	
	Sacramento	University of California-Davis, California State University at Davis, Hewlett-Packard, Signetics, Intel, Teledyne, Shugart	Sacramento Commerce and Trade Organization, Sacramento		×	
	San Diego: "Golden Triangle"	University of California-San Diego San Diego State University, Scripps Institute of Oceanography, General Dynamics, Rohr Industries	San Diego Economic Development Corporation, San Diego		×	
Maryland	Montgomery County: "Satellite Alley"	COMSAT, Fairchild, Litton, IBM, NASA, NSA, National Institute of Health	Maryland Industrial Development Board, Annapolis		×	

Appendix C (continued)

State	Area	Participants: Universities, Government Entities, Base Companies	Government Agency	Mature High-Technology Centers	Developing High-Technology Centers	Emerging High-Technology Centers
	Prince Georges County	University of Maryland-College Park, Litton, NASA, OAO, Martin Marietta	Prince Georges Economic Development Corp., Landover		×	
Massachusetts	Route 128, Boston	MIT, Harvard, Boston University, Tufts, Northeastern, DEC, Wang, Honeywell, GE, GTE, RCA, Raytheon	Massachusetts Department of Commerce and Development, Boston	×		
Florida	Orlando area: "Electronics Belt"	Pratt & Whitney, GE, IBM, Westinghouse, Honeywell, Harris Corp., Martin Marietta, Western Electric	Florida Division of Economic Development, Florida Department of Commerce, Tallahassee		×	
	Dade, Broward, Palm Beach counties: "Silicon Beach"	University of Miami	Florida Division of Economic Development, Florida Department of Commerce, Tallahassee		×	
	Gainsville to Orlando: "Robot Alley"	University of Florida-Gainsville, IBM, GE, Westinghouse	Florida Division of Economic Development, Florida Department of Commerce, Tallahassee			×
New York	Long Island: "Tech Island"	SUNY at Stony Brook, Polytechnic Institute of New York, Grumman Aerospace, Brookhaven National Laboratories, Cold Springs Harbor Laboratories, Harris Corp.	New York State Science & Technology Foundation, Albany		×	

Appendix C (continued)

	Syracuse	Syracuse University, Carrier, GE, Research Corporation of Syracuse, Niagara Scientific	New York Science & Technology Foundation, Albany		×
Texas	Austin and San Antonio	University of Texas-Austin, University of Texas-San Antonio, Motorola, Lockheed, Tandem	Texas Industrial Community, Austin	×	
	Dallas-Ft. Worth: I-20	University of Texas-Dallas, University of Dallas, Texas Instruments, E-Systems, Sunrise Systems, Nuclear Medicine Labs	Texas Industrial Community, Austin	×	
	Houston: I-610 and I-45 to Woodlands	Texas A&M, Rice, University of Houston, Texas Medical Center, Litton, Shamrock, Visidyne, Switch Data, NASA, oil companies	Texas Industrial Community, Austin	×	
New Mexico	Rio Grande Research Corridor	New Mexico Tech, University of New Mexico, New Mexico State University, Intel, Motorola, Signetics, GTE, GE, Western Electric, Kirkland AFB, Los Alamos Labs, Sandia Labs, Sperry Rand	New Mexico Economic Development Division, Santa Fe		
Virginia	Fairfax County: I-95 and Washington	George Mason University, ATT Long Lines, GTE, McDonnell-Douglas, Westinghouse	Fairfax County Economic Development Authority, Vienna	×	

Appendix C (continued)

State	Area	Participants: Universities, Government Entities, Base Companies	Government Agency	Mature High-Technology Centers	Developing High-Technology Centers	Emerging High-Technology Centers
Ohio	Cleveland	Lewis Research Center (NASA), Defense Contract Administration, Case Western Reserve University, Picker International, Johnson & Johnson, TRW, Bendix	Department of Economic Development, city of Cleveland, Cleveland		×	
	Columbus	Ohio State University, Western Electric, Bell Labs, Rockwell International, Battelle Memorial Research Institute	State Department of Development, Columbus		×	
	Cincinnati	University of Cincinnati, GE, Cincinnati Milicron, Structural Dynamic Research Corp.	Cincinnati Chamber of Commerce			×
	Dayton	University of Dayton, Wright State University, NCR, Mead, Wright-Patterson AFB, Air Force Institute of Technology, Monsanto Research, Bendix, Grumman	Dayton Development Council			×
Pennsylvania	Philadelphia: Route 202	University of Pennsylvania (Wharton), Drexel University, University City Science Center, IBM, Commodore	Technology Council, Chamber of Commerce, Philadelphia		×	

Appendix C (continued)

	Pittsburgh	Alcoa, Pittsburgh Plate Glass, US Steel, Westinghouse, Gulf, University of Pittsburgh, Carnegie-Mellon	Commonwealth of Pennsylvania, Department of Commerce, Harrisburg	×	
Washington	Seattle-Bellevue: I-5 corridor	University of Washington, Boeing, Eldec Corp., John Fluke Co., Squibb, Weyerhauser	Department of Commerce and Economic Development, Olympia	×	
Tennessee	Knoxville-Oak Ridge	University of Tennessee, Oak Ridge National Laboratories, Boeing, Goodyear Aerospace, Westinghouse, Magnavox	Tennessee Technical Foundation, Knoxville		×
New Jersey	Princeton	Princeton University, RCA, Grumman Aerospace, American Cyanamid, Exxon, Mobile	New Jersey Department of Commerce and Economic Development, Trenton	×	
Colorado	Colorado Springs	University of Colorado-Colorado Springs, Rolm, TRW, Ford Aerospace, Honeywell	Division of Commercial Development, state of Colorado, Denver	×	
	Denver-Boulder	University of Colorado-Boulder, Colorado State University, DEC, NCR, Hewlett-Packard	Division of Commercial Development, state of Colorado, Denver	×	
Illinois	Chicago	Northwestern University, University of Illinois, Illinois Institute of Technology, University of	Illinois Department of Commerce, Chicago	×	

Appendix C (continued)

State	Area	Participants: Universities, Government Entities, Base Companies	Government Agency	Mature High-Technology Centers	Developing High-Technology Centers	Emerging High-Technology Centers
		Chicago, Bell Labs, Western Electric, Amoco, Abbott Labs, Searle, Gould, Northrup, Fermi Labs, Argonne National Labs				
Alabama	Huntsville	University of Alabama-Huntsville, Redstone Arsenal, Intergraph Inc., Army Corps of Engineers, Army Missile Command, Lockheed, Rockwell, Boeing	Development Division of Chamber of Commerce, Huntsville		×	
Arizona	Phoenix-Tempe	Arizona State University, Motorola, Sperry Rand, ITT, Intel, Goodyear, Honeywell, IBM	Arizona Office of Economic Planning and Development		×	
	Tucson	IBM, Hughes Aircraft, Anaconda Copper, National Semiconductor, University of Arizona-Tucson	Tucson Economic Development Corp., Tucson		×	
Michigan	Ann Arbor	University of Michigan, Ford, GM, Chrysler, Bendix	Office of Economic Development, Department of Commerce, Lansing			×
Louisiana	Lafayette: "Silicon Bayou"	University of SW Louisiana, Regional Vocational Technical School, Celeron, Shell, Texaco, NASA, Exxon	Lafayette Harbor Terminal & Industrial Development District, Lafayette		×	

Appendix C (continued)

Minnesota	Minneapolis-St. Paul	University of Minnesota, 3M, Control Data, Honeywell, Cray Research	Minnesota High Tech Council, Minneapolis		×	
Utah	Salt Lake City	University of Utah, Eaton, UNIVAC Aerospace, US Steel, Kennecott Copper	Utah Economic Development Division, Salt Lake City		× ×	
North Carolina	Raleigh-Durham-Chapel Hill: "Research Triangle"	North Carolina State, University of North Carolina, Duke, IBM, Environmental Protection Agency, Becton, Dickenson, GE Semiconductor, Burroughs, Data General, Northern Telecom	North Carolina Department of Commerce Industrial Development Division, Raleigh	×		
Georgia	Atlanta	Georgia Tech, Rockwell, Scientific Atlanta	Office of the governor, Atlanta		×	
Rhode Island	Newport, Portsmouth, Middletown: Aquidneck Island	Naval War College, Brown University, University of Rhode Island, Raytheon Submarine Division, U.S. Navy Underwater Systems Center, Gould, Goodyear	Rhode Island Department of Economic Development, Providence			×
Indiana	Indianapolis	Purdue, Indiana University, GM, Eli Lilly, Renault, International Harvester, Naval Avionics Center	Office of the mayor, Indianapolis		×	
Wisconsin	Madison	University of Wisconsin-Madison, University of Madison Hospital, GE Medical Systems,	Wisconsin Department of Development, Madison			×

Appendix C (continued)

State	Area	Participants: Universities, Government Entities, Base Companies	Government Agency	Mature High-Technology Centers	Developing High-Technology Centers	Emerging High-Technology Centers
		Ohio Medical Labs, Nicolet Instruments, Cray Research				
Oregon	Tualatin Valley: "Sunset Corridor" west of Portland	Tektronix, Intel	Business and Community Development Department, state of Oregon, Salem		×	
	Willamette Valley: I-5 Portland to Eugene	Oregon State University, Hewlett-Packard, Spectra Physics	Business and Community Development Department, state of Oregon, Salem		×	
	Bend-Richmond	Bend Research	Economic Development Department, Salem			×
South Carolina	Columbia	Monsanto, GE, Sony, United Technologies, NCR, DEC	State Development Board of South Carolina, Columbia			×
Oklahoma	Entire state	Western Electric, GM, oil companies, University of Oklahoma-Norman, Oklahoma State University, Tinker AFB	State Office of Economic Development, Oklahoma City			×
New Hampshire	Salem-Manchester-Nashua:	University of New Hampshire, Lowell University,	New Hampshire Office of Industrial Development,			

Appendix C (continued)

	"Golden Triangle"	DEC, Bedford Computer, Sanders Association, Kollsman Instruments, Computer-Vision, Data General	Concord	×	
Arkansas	Little Rock to Pine Bluff: Technology Corridor	University of Arkansas-Pine Bluff and Little Rock, Little Rock Medical Center, BEI Electronics, Pine Bluff Arsenal, National Center for Toxicological Research	Arkansas Industrial Development Commission, Little Rock		×
Maine	Portland	University of Southern Maine, Data General, DEC, Fairchild Semiconductor, Sprague Electric	Maine State Development Office, Augusta		×
Vermont	Burlington	University of Vermont, GE, IBM, DEC, McDonnell-Douglas, Bendix	State of Vermont Economic Development Department, Montpelier		×

Source: *Venture*, September 1983, with permission from *Venture Magazine*, Inc., New York.

D
Selected University/Corporation Programs

Business Firm	Activity	Funding	Academic Institution
Monsanto Co.	Biomedical proteins and peptides regulate cellular functions - 30% basic, 70% applied to human diseases	\$23.5 million; 5 years, renewable	Washington University, St. Louis, Mo.
Monsanto Co.	Will sponsor Basic Research in Plant Molecular Biology structure and regulation of plant genes	\$4 million; 5 years	Rockefeller University
IBM	Develop manufacturing engineering courses; 1981 IBM grants totaled \$17 million	\$50 million; \$10 million in cash, \$40 million in equipment	Five universities share \$10 million cash (to be announced); 20 universities receive the equipment; includes University of Texas-Austin
Green Cross Corp. (OSAKA)	Mass producing monoclonal antibodies by all fusion techniques to combat cancer	\$ unknown, 2-year contract signed	University of California
American Cyanamid-Lederle Labs	Pathway to generate chemical mediators causing allergic reactions to develop drugs to block released mediators	\$2.5 million; 5-year grant	Johns Hopkins School of Medicine
W. R. Grace Co.	Research in microbiology	\$6–8 million; 5-year grant	Massachusetts Institute of Technology
Apple Computer to Xerox (26 companies)	Microelectronic innovations; 31 high-technology research projects	\$2.2 million in cash and equipment	University of California Microelectronic Innovations and Computer Research Opportunities Program, 6 University of California campuses
IBM	Robotics and use of computers and assembly lines	\$1 million grant	University of Pennsylvania School of Engineering and Applied Science

Appendix D (continued)

NSF and coalition of some 30 industrial companies	Establish the University/Industry Cooperative Center for Robotics	$ unknown	Site: University of Rhode Island
Celanese Corp.	Specific basic biotechnology research	$1.1 million; 3-year term	Yale University
Bristol-Meyers Co.	Developing anticancer drugs, company option to license cancer chemotherapy drugs discovered by participating Yale faculty	$3 million; 5-year cooperation agreement	Yale University
Gould Inc.	Gould Lab computer service facility	$500,000 over next 5 years	Brown University
IBM	NYU Robotics Center, math, geometric molding, and software	Major contribution from IBM and equipment value unknown	New York University
NSF grant plus Carolina Power and Lights, Digital Equipment, Exxon, General Telephone & Electronic, IBM, ITT, Western Union and Western Electric	North Carolina State's University/Industry Cooperative Research Center for Communications and Signal Processing; basic and applied research	NSF, $650,000; 5-year grant Industrial sponsors, $50,000 each for first 5 years	North Carolina University
Hoechst	Biotechnology research	$70 million over 10-year period	Massachusetts General Hospital and Harvard University
Dupont	Genetic engineering	$6 million over 5 years	Harvard Medical School
Monsanto	Tumor angiogenesis factor	$23 million over 12 years	Harvard University
Engenics (consists of Bendix, General Foods, Koppers, Mean, MacLaren, and Elf Technologies)	Industrial microbiology	$1 million; 4 years	University of California at Berkeley and Stanford
Syntex & Hewlett-Packard	Biotechnology	$600,000 per year for 3 years	Stanford University
Exxon	Combustion research	$7–8 million; 10 years	MIT

D (continued)

Business Firm	Activity	Funding	Academic Institution
Westinghouse	Robotics	$1.2 million per year	Carnegie-Mellon
Industry Participants	Industry scientists work for a year at CalTech and get view of ongoing research and share expertise with faculty and staff	$100,000 each	CalTech
IBM, General Electric, and Norton	Research funds and equipment for a Center for Integrated Structures	So far: $1.25 million from GE for 3 years; Norton Co. donated building; and IBM provided a $2.75 million electron beam lithograph system	Rensselaer Polytechnic Institute
Consortium Caterpillar Tractor Co. Cummins Engine Co. John Deere Co. United Technologies Research Center	Engine research includes diesel engines and fuel	$ unknown	MIT-Sloan Automotive Labs
MCC	MCC in Austin: long-range programs 1. Cost-effective interconnection of computers using VLSI chips + $1 million + circuit elements 2. 8 to 10-year advanced computer architecture study 3. Breakthroughs in CAD/CAM systems 4. Quantum improvement in procedures and tools centered on expert and knowledge-based system	MCC budget after startup $50–100 million per year	University of Texas System at Austin and Texas A&M University

Source: IC^2, the University of Texas at Austin, 1983.

E
Selected Data

Eugene B. Konecci

TABLE E.1
ASSESSMENT OF PUBLIC FACILITIES: THEIR CONDITION AND TYPE OF WORK NEEDED

	Type of Work Needed (%)			
Type of Facility	Good	Repair	Rehabilitate	Replace
Status: More than 50% Good				
Community social service facilities[a]	58.8	10.4	6.6	7.5
Parks, recreation facilities, open space	55.7	18.4	14.0	2.7
Water storage	54.8	7.5	8.0	3.2
Public buildings[b]	54.4	16.2	12.1	14.6
Water treatment	50.6	5.7	6.7	4.2
Traffic control equipment	50.3	16.6	11.2	11.2
Status: Less Than 50% Good				
Streets and roads	25.8	38.7	25.8	6.0
Sidewalks and curbs	25.3	38.4	22.7	8.8
Stormwater collection, drainage	23.9	24.4	31.9	11.1
Sewage collection	36.2	16.6	28.3	4.9
Bridges, overpasses, viaducts	2.5	19.0	13.3	12.5
Wastewater treatment	37.2	9.9	17.4	7.4
Water distribution	46.4	13.5	14.0	3.9

[a]Community social service facilities include senior citizens centers, libraries, and municipal cultural centers.

[b]Public buildings include city hall, municipal garages, and fire and police facilities.

Source: National League of Cities (NLC) and U.S. Conference of Mayors (USCM) Infrastructure Survey, 1983.

TABLE E.2
SURVEY RESULTS OF MUNICIPALITIES' ABILITIES TO FINANCE PUBLIC FACILITIES

Type of Facility	Could Use Own Resources (%)	Need Help (%)	Isn't in Budget (%)
Public buildings	67.8	17.4	1.1
Sidewalks, curbs	63.3	31.4	2.3
Water distribution	57.8	17.8	15.1
Water storage	53.4	18.2	17.3
Traffic control equipment	50.3	38.6	6.2
Streets, roads	35.0	62.0	0.6
Wastewater treatment	16.7	54.5	20.8
Bridges, overpasses, viaducts	13.8	54.4	19.0
Parks, recreation facilities, open space	47.0	42.9	4.1
Stormwater collection, drainage	41.8	48.3	5.0
Sewage collection	41.3	44.1	9.6
Public hospitals, clinics	6.5	7.3	60.8
Docks, wharves, ports	5.1	8.2	57.7
Public school buildings	13.7	10.9	53.8
Public transportation facilities	4.9	22.7	51.1
Public transportation rolling stock	5.7	24.0	48.9
Solid waste disposal, resource recovery facilities	33.2	23.8	29.2

Source: National League of Cities (NLC) and U.S. Conference of Mayors (USCM) Infrastructure Survey, 1983.

TABLE E.3
ESTIMATED 1983–90 CAPITAL EXPENDITURES FOR INFRASTRUCTURE SYSTEM

A. Capital Spending for Infrastructure
($ billion per year, 1982 dollars)

Infrastructure System	Estimated Annual Needs 1983–90 Total	New	Repair	Federal Share
Highways	$27.2	9.9	17.3	13.1
Public transit	5.5	2.2	3.3	4.1
Wastewater treatment	6.6	6.1	0.5	4.2
Water resources (Ports, dams, etc.)	4.1	2.3	1.8	3.7
Air traffic control	0.8	0.1	0.7	0.8
Airports	1.5	1.0	0.5	0.9
Municipal water supply	7.7	3.6	4.1	1.4
Total	$53.4	25.2	28.2	28.2

B. Estimated 1983–90 Capital Expenditures for
Infrastructure Systems (in percent)

Infrastructure Component	Total	New	Repair	Federal Share
Highways	50.9	18.5	32.4	24.5
Municipal water supply	14.4	6.7	7.7	2.6
Wastewater treatment	12.4	11.5	0.9	7.9
Public transit	10.3	4.1	6.2	7.7
Water resources (ports, dams, etc.)	7.7	4.3	3.4	6.9
Airports	2.8	1.9	0.9	1.7
Air traffic control	1.5	0.2	1.3	1.5
Totals	100.0	47.2	52.8	52.8

C. Federal Share of Annual Capital Costs
($ billion)

Infrastructure System	Current Spending Level	Under Current Programs	Under Revised Programs
Highways	$12.7	13.1	9.3
Public transit	3.7	4.1	2.2
Wastewater treatment	3.2	4.2	3.7
Water resources	2.3	3.7	3.1
Air traffic control	0.8	0.8	0.7
Airports	0.8	0.9	0.3
Municipal water supply	0.9	1.4	1.0
Total	$24.4	28.2	20.3

Source: *Public Works Infrastructure: Policy Considerations for the 1980s*, Congressional Budget Office, Washington D.C. 1983.

TABLE E.4
POSSIBLE ACTIONS FOR STATE GOVERNMENTS

Action	Suitable States	Benefits	Drawbacks
Improve technical assistance programs	All	Facilitate bond issuance and encourage responsible debt management	Local governments may fear state intrusion
Create a loan program for water and sewer construction	All	Stimulate investment in water and sewer facilities, and supplement federal wastewater program; provide loans at favorable rates, particularly for distressed communities and small or infrequent issuers	Increase state general obligation debt and thereby lower state credit rating
Create a municipal bond bank	Rural states with many small issuers	Reduce borrowing costs for small or infrequent issuers	Local banks, bond counselors, and underwriters may suffer a loss in business
Earmark state aid for debt service	States with fiscally distressed communities	Improve rating of bond issues (reduce interest rate) for large, fiscally distressed communities	Requires a large permanent state aid program and some state supervision
Assist local governments with creative financing (through technical assistance programs and enabling legislation)	All	Increase local flexibility, facilitate use of beneficial techniques, and discourage improper use	Many techniques are untested; may result in excessive short-term debt, or high interest rates in future
Create loan programs for energy impact assistance	Energy-rich states	Finances rapid capital construction energy boom towns; supplement to grant programs	Local governments prefer grants or sharing of severance revenues

TABLE E.4 (continued)

Increase state supervision of local debt management	States with cities having poor credit ratings or histories of poor financial management	Encourage responsible debt management, and improve credit ratings	Increased administrative costs for local governments, possible restriction on local actions
Guarantee local debt	Most states	Improve rating on local bonds	May seriously weaken state credit rating

Note: The most effective measures are listed first.
Source: National Conference of State Legislatures, Denver, Colorado: Richard Watson (1982).

TABLE E.5
FEDERAL OUTLAYS FOR DEFENSE, NASA AND AEROSPACE PRODUCTS AND SERVICES: FISCAL YEARS 1960–84 (in millions of dollars)

Year	Total National Defense	Total NASA	Federal Outlays for Aerospace Products and Services			Aerospace as Percentage of Total National Defense and NASA
			Total	DOD[a]	NASA	
1960	$ 45,691	$ 401	$12,849	$12,502	$ 347	27.9
1961	47,494	744	13,606	12,960	646	28.2
1962	51,103	1,257	15,135	13,992	1,143	28.9
1963	52,755	2,552	16,186	13,857	2,327	29.3
1964	53,591	4,171	17,938	14,205	3,733	31.1
1965	49,578	5,093	15,697	11,135	4,561	28.7
1966	56,785	5,933	17,771	12,411	5,360	28.3
1967	70,081	5,426	20,011	14,874	5,137	26.5
1968	80,517	4,724	21,355	16,757	4,598	25.1
1969	81,232	4,251	20,472	16,286	4,185	23.9
1970	80,295	3,753	18,747	15,046	3,699	22.3
1971	77,661	3,382	17,335	13,997	3,338	21.4
1972	78,336	3,422	16,999	13,627	3,372	20.8
1973	74,571r	3,315	15,945	12,675	3,270	20.5
1974	77,781r	3,256	15,782	12,601	3,181	19.5
1975	85,552r	3,266	15,943	12,762	3,181	18.0
1976	89,430r	3,669	16,843	13,295	3,548	18.1
Tr. Qtr.	22,307r	952	3,944	3,018	926	17.0
1977	97,501	3,945	18,201	14,361	3,840	17.9
1978	105,186	3,983	12,624	8,765	3,859	11.6
1979	117,681	4,196	14,984	10,920	4,064	12.3
1980	135,856	4,852	18,297	13,585	4,712	13.0
1981	159,765	5,426	21,984	16,706	5,278	13.3
1982	187,418	6,035	27,057	21,131	5,926	14.0
E1983	214,769	6,722	33,816	27,231	6,585	15.3
E1984	245,305	6,981	41,346	34,494	6,852	16.4

[a]Prior to 1978, DOD outlays for aircraft and missile procurement and RDT&E. Effective 1978, includes only procurement; outlays for RDT&E by product group not available.

r, Revised; E, estimated.

Note: "National Defense" includes the military budget of the Department of Defense and other defense-related activities. "Total NASA" includes R&D activities, administrative operations and construction of facilities. NASA construction is not included in "Total Aerospace Products and Services."

Source: *The Budget of the United States Government* (annual 1960–83). Washington D.C.: Government Printing Office.

TABLE E.6

FEDERAL OUTLAYS FOR AEROSPACE PRODUCTS AND SERVICES: FISCAL YEARS 1960–84 (in millions of dollars)

Year	TOTAL	Department of Defense[a]			NASA[b]
		Total	Aircraft	Missiles	
1960	$12,849	$12,502	$ 7,416	$5,086	$ 347
1961	13,606	12,960	6,963	5,997	646
1962	15,135	13,992	7,773	6,219	1,143
1963	16,186	13,857	7,799	6,058	2,327
1964	17,938	14,205	8,276	5,929	3,733
1965	15,697	11,135	7,138	3,997	4,562
1966	17,771	12,411	8,541	3,870	5,360
1967	20,011	14,874	10,442	4,432	5,137
1968	21,355	16,757	12,016	4,741	4,598
1969	20,472	16,286	11,367	4,919	4,185
1970	18,747	15,048	9,940	5,108	3,699
1971	17,335	13,997	8,849	5,148	3,338
1972	16,999	13,627	8,461	5,166	3,372
1973	15,945	12,675	7,614	5,061	3,270
1974	15,782	12,601	7,460	5,141	3,181
1975	15,943	12,762	7,697	5,065	3,181
1976	16,843	13,295	8,704	4,591	3,548
Tr. Qtr.	3,944	3,018	2,096	922	926
1977	18,201	14,361	9,321	5,040	3,840
1978	12,624	8,765	6,971	1,794	3,859
1979	14,984	10,920	8,836	2,084	4,064
1980	18,297	13,585	11,124	2,461	4,712
1981	21,984	16,706	13,193	3,513	5,278
1982	27,057	24,131	16,793	4,338	5,926
E1983	33,816	27,231	21,436	5,795	6,585
E1984	41,346	34,494	27,066	7,428	6,852

[a]Prior to 1978, DOD outlays for aircraft and missile procurement and RDT&E. Effective 1978, includes only procurement; outlays for RDT&E by product group not available.

[b]Includes Research & Development Program Management, excludes Construction of Facilities.

E, Estimate.

Source: *The Budget of the United States Government* (annual 1960–83). Washington D.C.: Government Printing Office.

TABLE E.7
NATIONAL AERONAUTICS AND SPACE ADMINISTRATION OUTLAYS: FISCAL YEARS 1960–84 (in millions of dollars)

Year	TOTAL	Research and Development	Construction of Facilities	Research and Program Management
1960	$ 401	$ 256	$ 54	$ 91
1961	744	487	98	159
1962	1,257	936	114	207
1963	2,552	1,912	225	416
1964	4,171	3,317	438	416
1965	5,093	3,984	531	578
1966	5,933	4,741	573	619
1967	5,426	4,487	289	650
1968	4,724	3,946	126	652
1969	4,251	3,530	65	656
1970	3,753	2,992	54	707
1971	3,382	2,630	44	708
1972	3,422	2,623	50	749
1973	3,315	2,541	45	729
1974	3,256	2,421	75	760
1975	3,266	2,420	85	761
1976	3,669	2,749	121	799
Tr. Qtr.	952	731	26	195
1977	3,945	2,980	105	860
1978	3,983	2,989	124	870
1979	4,196	3,139	133	925
1980	4,852	3,702	140	1,010
1981	5,426	4,228	147	1,050
1982	6,035	4,796	109	1,130
E1983	6,722	5,335	137	1,250
E1984	6,981	5,605	129	1,248

E, Estimate.
Note: Detail may not add to totals because of rounding.
Source: *The Budget of the United States Government* (annual 1960–83). Washington D.C.: Government Printing Office.

TABLE E.8
SPACE ACTIVITIES BUDGET AUTHORITY:
FISCAL YEARS 1959-83 (in millions of current dollars)

Year	TOTAL	NASA[a]	DOD	Energy	Commerce	Other[b]
1959	$ 785	$ 261	$ 490	$ 34	$ —[c]	$—
1960	1,066	462	561	43	—	( )[d]
1961	1,808	926	814	68	—	1
1962	3,295	1,797	1,298	148	51	1
1963	5,435	3,626	1,550	214	43	2
1964	6,831	5,016	1,599	210	3	3
1965	6,956	5,138	1,574	229	12	3
1966	6,970	5,065	1,689	187	27	3
1967	6,710r	4,830	1,664	184	29	3
1968	6,529r	4,430	1,922	145	28	4
1969	5,976	3,822	2,013	118	20	3
1970	5,341	3,547	1,678	103	8	4
1971	4,741	3,101	1,512	95	27	5
1972	4,575	3,071	1,407	55	31	10
1973	4,825	3,093	1,623	54	40	15
1974	4,640	2,759	1,766	42	60	14
1975	4,914	2,915	1,892	30	64	13
1976	5,320	3,225	1,983	23	72	16
Tr. Qtr.	1,341	849	460	5	22	4
1977	5,983	3,440	2,412	22	91	18
1978	6,509	3,623	2,729	34	103	20
1979	7,419	4,030	3,211	59	98	21
1980	8,689	4,680	3,848	40	93	28
1981	9,978	4,992	4,828	41	87	30
E1982	12,041	5,462	6,387	38	125	29
E1983	14,839	6,122	8,502	31	151	34

[a]Excludes amounts for air transportation.
[b]Department of Interior and Agriculture and the National Science Foundation.
[c]Data not available.
[d]Less than $500,000.
r, Revised; E, estimate.
Note: Detail may not add to totals because of rounding.
Source: NASA, *Aeronautics and Space Report of the President* (annual 1960-83). National Aeronautics and Space Administration, Washington, D.C.

TABLE E.9

SPACE ACTIVITIES BUDGET AUTHORITY IN CONSTANT DOLLARS: FISCAL YEARS 1959–83 (in millions of constant dollars, 1972 = 100[a])

Year	Total	NASA[b]	DOD	Energy	Commerce	Other[c]
1959	$1,152	$ 283	$ 719	$ 50	$ —[d]	$—
1960	1,537	666	809	62	—	( )[c]
1961	2,577	1,320	1,160	97	—	1
1962	4,616	2,518	1,818	207	71	1
1963	7,341	4,897	2,093	289	58	3
1964	9,226	6,775	2,160	284	4	4
1965	9,277	6,852	2,099	305	16	4
1966	9,055	6,580	2,194	243	35	4
1967	8,443	6,078	2,094	232	36	4
1968	7,930	5,381	2,335	176	34	5
1969	6,935	4,435	2,336	137	23	3
1970	5,866	3,896	1,843	113	9	4
1971	4,958	3,243	1,581	99	28	5
1972	4,575	3,071	1,407	55	31	10
1973	4,620	2,962	1,554	52	38	14
1974	4,145	2,464	1,577	38	54	13
1975	3,993	2,369	1,537	24	52	11
1976	4,041	2,450	1,506	17	55	12
Tr. Qtr.	991	627	340	4	16	3
1977	4,257	2,447	1,716	16	65	13
1978	4,340	2,416	1,820	23	69	13
1979	4,555	2,474	1,972	36	60	13
1980	4,916	2,648	2,177	23	53	16
1981	5,113	2,558	2,474	21	45	15
E1982	5,820	2,640	3,087	18	60	14
E1983[c]	6,942	2,864	3,977	15	71	16

[a]Based on fiscal year GNP implicity price deflator.

[b]Excludes amounts for air transportation.

[c]Departments of Interior and Agriculture and the National Science Foundation.

[d]Data not available.

[e]Based on average of 1983 first- and second-quarter GNP implicit price deflator.

E, estimate.

Note: Detail may not add to totals because of rounding.

Source: NASA, *Aeronautics and Space Report of the President* (annual 1960–83). National Aeronautics and Space Administration, Washington, D.C.

TABLE E.10

NET PROFIT AFTER TAXES AS A PERCENTAGE OF SALES, ASSETS, AND EQUITY FOR ALL MANUFACTURING CORPORATIONS AND THE AEROSPACE INDUSTRY: CALENDAR YEARS 1968-83

As a Percentage of Sales

Year	All Manufacturing Corporations	Nondurable Goods	Durable Goods	Aerospace Industry[a]
1968	5.1	5.3	4.9	3.2
1969	4.8	5.0	4.6	3.0
1970	4.0	4.5	3.6	2.0
1971	4.1	4.5	3.8	1.8
1972	4.4	4.6	4.3	2.4
1973	4.7	5.0	4.5	2.9
1974	5.5	6.4	4.7	2.9
1975	4.6	5.1	4.1	2.9
1976	5.4	5.5	5.2	3.4
1977	5.3	5.3	5.3	4.2
1978	5.4	5.4	5.5	4.4
1979	5.7	6.1	5.2	5.0
1980r	4.8	5.6	4.0	4.3
1981	4.7	5.2	4.2	4.3
1982				
1st quarter	3.8	4.4	3.0	3.2
2nd quarter	3.8	4.2	3.4	3.1
3rd quarter	3.5	4.4	2.4	3.8
4th quarter	2.8	4.5	0.7	3.1
Average	3.5	4.4	2.4	3.3
1983				
1st quarter	3.2	4.0	2.1	3.2
2nd quarter	4.2	4.8	3.4	4.1

As a Percentage of Assets[b] and Equity[b]

	All Manufacturing Corporations		Aerospace Industry[a]	
Year	Percentage of Assets	Percentage of Equity	Percentage of Assets	Percentage of Equity
1968	6.6	12.1	4.4	14.2
1969	6.1	11.5	3.5	10.6
1970	4.9	9.3	2.2	6.8
1971	5.1	9.7	2.0	5.8
1972	5.5	11.1	2.7	8.6
1973	6.5	12.8	2.4	10.3
1974	8.0	14.9	3.7	10.4
1975	6.2	11.6	3.8	11.0
1976	7.5	14.0	4.7	12.8
1977	7.6	14.2	5.7	14.9
1978	7.8	15.0	5.5	15.7
1979	8.4	16.5	6.3	18.4
1980r	6.9	13.9	5.2	16.0
1981	6.7	13.6	5.0	15.6
1982				
1st quarter	4.83	10.06	3.39	10.99
2nd quarter	5.06	10.48	3.62	11.66
3rd quarter	4.43	9.20	4.11	13.36
4th quarter	3.50	7.21	3.71	11.99
Average	4.46	9.24	3.71	12.00
1983				
1st quarter	3.87	7.99	3.74	11.99
2nd quarter	5.37	11.05	5.01	15.43

[a]Based on a sample of corporation entities classified in SIC codes 372 and 376, having as their principal activity the manufacture of aircraft, guided missiles, space vehicles, and parts.

[b]Average of four quarters.

r, Revised.

Source: Federal Trade Commission, *Quarterly Financial Report for Manufacturing, Mining and Trade Corporations.* Washington D.C. 1983.

TABLE E.11
INCOME ACCOUNTS OF AEROSPACE COMPANIES:
CALENDAR YEARS 1977–83 (in millions of dollars)

	1977	1978	1979	1980r	1981	1982	1983[a]
Net Sales	$34,307	$41,689	$51,801	$60,638	$67,341	$66,198	$36,376
Income from operations	2,338	3,023	3,606	3,659	3,735	3,090	1,918
Total income before income taxes	2,296	2,726	3,711	3,647	4,518	3,272	2,213
Provision for federal income taxes	1,003	1,154	1,489	1,341	1,641	1,081	799
As a percentage of total income	43.7	42.3	40.1	36.8	36.3	33.0	37.6
Net profit after taxes	1,427	1,816	2,614	2,588	2,877	2,193	1,324
As a percentage of net sales	4.2	4.4	5.0	4.3	4.3	3.3	3.6
Net profit retained in business	1,012	1,255	1,897	1,790	1,985	1,350	878

[a]First two quarters of 1983.
r, Revised.
Note: Based on sample of corporate entities classified in SIC codes 372 and 376, having as their principal activity the manufacture of aircraft, guided missiles, space vehicles, and parts.
Source: Federal Trade Commission, *Quarterly Financial Report for Manufacturing, Mining and Trade Corporations.* Washington D.C. 1983.

TABLE E.12

SALES OF MAJOR AEROSPACE COMPANIES AS REPORTED BY THE BUREAU OF THE CENSUS: CALENDAR YEARS 1968–83 (in millions of current dollars)

Year	Grand Total	Total		Aircraft, Engines, and Parts		Missiles and Space Includes Propulsion	Other Aerospace		Nonaerospace
		U.S. Government	Other	U.S. Government	Other		U.S. Government	Other	
1968	$25,592	$16,635	$ 8,957	$ 7,411	$ 6,439	$ 6,076	$ 2,077	$1,040	$ 2,549
1969	24,648	16,560	8,088	7,161	5,603	5,660	2,539	986	2,699
1970	24,752	16,407	8,345	7,586	5,880	5,422	2,324	896	2,644
1971	21,679	14,114	7,565	6,313	5,079	4,971	1,909	884	2,523
1972	21,499	13,492	8,007	4,954	5,199	5,598	2,067	1,035	2,646
1973	24,305	14,431	9,874	5,539	6,739	5,580	2,103	1,001	3,343
1974	26,849	15,196	11,653	5,982	7,560	5,854	2,101	1,285	4,067
1975	29,473	17,314	12,159	6,859	7,797	6,310	2,070	1,645	4,792
1976	31,328	19,083	12,245	8,314	7,622	5,880	2,368	1,833	5,311
1977	33,315	20,704	12,611	8,848	7,530	5,775	2,839	2,219	6,104
1978	37,963	21,888	16,080	8,724	10,581	6,380	3,363	2,107	6,813
1979	46,173	23,229	22,944	8,649	16,023	7,197	3,930	2,659	7,715
1980	58,440	26,674	31,766	9,427	20,097	8,393	6,869	2,609	11,045
1981	70,536	32,504	38,032	12,168	22,527	9,842	8,170	3,120	14,709
1982									
1st quarter	16,719	8,740	7,979	1,869	5,428	2,705	2,243	803	3,671
2nd quarter	18,113	9,317	8,796	2,014	5,764	2,981	2,261	986	4,106
3rd quarter	18,869	11,456	7,413	3,703	4,588	2,938	2,644	1,103	3,893
4th quarter	20,377	12,086	8,309	3,550	5,342	3,144	3,098	815	4,128
Total	74,078	41,599	32,497	11,136	21,122	11,768	10,246	3,707	15,798
1983									
1st quarter	19,684	10,849	8,835	3,097	6,401	2,979	2,615	794	3,798

Source: Bureau of the Census, *Current Industrial Reports*, Series MQ37D (Quarterly). Washington D.C. 1983.

TABLE E.13

SALES OF MAJOR AEROSPACE COMPANIES AS REPORTED BY THE BUREAU OF THE CENSUS: CALENDAR YEARS 1968–83 (in millions of constant dollars, 1972 = 100[a])

Year	Grand Total	Total: U.S. Government	Total: Other	Aircraft, Engines, and Parts: U.S. Government	Aircraft, Engines, and Parts: Other	Missiles and Space Includes Propulsion	Other Aerospace: U.S. Government	Other Aerospace: Other	Nonaerospace
1968	$31,006	$20,154	$10,852	$ 8,979	$ 7,801	$ 7,361	$ 2,516	$1,260	$ 3,088
1969	28,400	19,081	9,319	8,251	6,456	6,521	2,925	1,136	3,110
1970	27,066	17,941	9,125	8,295	6,430	5,929	2,541	980	2,891
1971	22,580	14,701	7,879	6,575	5,290	5,178	1,988	921	2,628
1972	21,499	13,492	8,007	4,954	5,199	5,598	2,067	1,035	2,646
1973	22,996	13,654	9,342	5,241	6,376	5,280	1,990	947	3,163
1974	23,363	13,228	10,140	5,205	8,578	5,094	1,828	1,118	3,539
1975	23,473	13,789	9,684	5,463	6,210	5,025	1,649	1,310	3,817
1976	23,714	14,445	9,269	6,293	5,769	4,451	1,792	1,388	4,020
1977	23,825	14,807	9,019	6,328	5,385	4,130	2,030	1,587	4,365
1978	25,304	14,587	10,716	5,814	7,052	4,252	2,241	1,404	4,540
1979	28,367	14,271	14,096	5,314	9,844	4,422	2,414	1,634	4,740
1980	32,950	15,039	17,910	5,315	11,331	4,732	3,873	1,471	6,227
1981	36,413	16,780	19,633	6,282	11,629	5,081	4,218	1,611	7,593
1982									
1st quarter	7,960	4,300	3,926	920	2,671	1,331	1,104	395	1,806
2nd quarter	8,797	4,520	4,267	977	2,796	1,446	1,097	478	1,992
3rd quarter	9,070	5,507	3,563	1,780	2,205	1,412	1,259	530	1,871
4th quarter	9,703	5,755	3,957	1,691	2,544	1,497	1,475	388	1,966
Total	35,809	20,109	15,709	5,383	10,210	5,689	4,953	1,792	7,637
1983									
1st quarter	9,250	5,098	4,152	1,455	3,008	1,400	1,229	373	1,785

[a]Based on GNP implicit price deflator.

Source: Bureau of the Census, *Current Industrial Reports*, Series MQ37D (Quarterly). Washington D.C. 1983.

TABLE E.14
BALANCE SHEET OF AEROSPACE COMPANIES: DECEMBER 31, 1977–83 (in millions of dollars)

	1977	1978	1979	1980	1981	1982	1983[a]
Assets							
Current Assets							
Cash[b]	$ 2,136	$ 2,696	$ 3,001	$ 562	$ 1,043	$ 1,746	$2,366
U.S. government securities	31	119	79				
Other securities/ commercial paper[b]	1,097	1,077	564	2,250	2,777	3,163	5,166
Total cash and U.S. government securities	$ 3,267	$ 3,894	$ 3,645	$ 2,812	$ 3,820	$ 4,910	$ 7,483
Receivables (total)	3,564	4,475	5,237	5,991	5,932	7,189	11,514
Inventories (gross)	10,568	15,968	20,491	26,497	29,966	11,560	10,981
Other current assets	677	840	844	834	870	3,528	4,066
Total current assets	$18,075	$25,195	$30,217	$36,135	$40,588	$27,186	$34,044
Net plant, property, and equipment	4,320	5,639	7,261	9,368	10,909	29,210	28,585
Other non-current assets	3,705	5,144	7,041	6,935	7,400	22,622	21,890
Total assets	$26,100	$35,978	$44,518	$52,437	$58,896	$79,018	$84,519
Liabilities							
Current liabilities							
Short-term loans	$ 279	$ 171	$ 698	$ 1,198	$ 1,701	$ 2,968	$ 1,935
Advances by U.S. government	1,886	5,400	6,554	( )[c]	( )[c]	( )[c]	( )[c]
Trade accounts and notes payable	2,757	3,296	4,266	5,095	5,193	7,708	9,967

TABLE E.14 (continued)

	1977	1978	1979	1980	1981	1982	1983[a]
Income taxes accrued	1,779	2,086	2,742	2,769	2,521	434	766
Installments due on long-term debts	307	249	272	178	279	808	1,595
Other current liabilities	4,612	7,940	9,342	19,589	22,072	12,378	14,376
Total current liabilities	$11,621	$19,144	$23,873	$28,830	$31,767	$24,296	$28,640
Long-term debt	4,117	3,637	3,975	4,525	5,347	9,453	7,504
Other noncurrent liabilities	496	1,016	1,356	2,123	2,923	9,946	10,387
Total liabilities	$16,233	$23,798	$29,204	$35,478	$40,036	$43,695	$46,531
Stockholders' equity							
Capitol stock	$ 3,452	$ 3,864	$ 5,013	$5,072	$ 5,622	$ 8,790	$10,215
Retained earnings	6,415	8,315	10,301	11,888	13,239	26,533	27,778
Total net worth	$ 9,866	$12,180	$15,315	16,959	$18,860	$35,323	$37,988
Total liabilities and stockholders' equity	$26,100	$35,978	$44,518	$52,437	58,896	$79,018	$84,519
Net working capitol	$ 6,454	$ 6,051	$ 6,344	$ 7,304	$ 8,822	$ 2,890	$ 5,404

[a]June 30, 1983.

[b]Effective 1980, deposits outside United States includes in "Other securities and commercial paper," they previously were included in "Other current liabilities."

[c]Included in "Other current liabilities."

Note: Based on sample of corporate entities classified in SIC codes 372 and 376, having as their principal activity the manufacture of aircraft, guided missiles, space vehicles, and parts.

Source: Federal Trade Commission, *Quarterly Financial Report for Manufacturing, Mining and Trade Corporations.* Washington D.C. 1983.

TABLE E.15
NEW JOBS IN HIGH-TECHNOLOGY INDUSTRIES CREATED BY INCREASED DEFENSE EXPENDITURES (Texas)

High-Technology Industry	1983	1984	1985	1986	1987
Computers	735	818	1,119	954	651
Electrical instruments	80	24	111	93	63
Electronic communication equipment	2,858	3,120	4,227	3,583	2,421
Electrical equipment	100	110	147	123	81
Aircraft	6,014	6,535	8,938	7,548	5,012
Scientific instruments	763	807	1,071	883	603
Mechanical measuring devices	693	746	1,003	837	557
Medical, photo, and optical instruments	46	50	66	56	36
Ordnance	1,536	1,658	2,221	2,030	1,291
Computer/accounting services	179	189	251	210	140
Total high technology	13,007	14,115	19,154	16,317	10,855
Percentage high technology	34	35	35	36	37
Total employment (increase)	38,365	40,374	53,259	44,871	28,993

Source: IC^2 Institute, the University of Texas at Austin, 1983.

TABLE E.16
PROJECTED INCREMENTAL LABOR AVAILABILITY (unemployment plus change in the labor force excluding agriculture and government) (in thousands of persons)

	1983	1984	1985	1986	1987
Mining	65.6	70.9	74.3	71.3	71.7
Construction	75.2	92.6	98.3	95.0	90.4
Manufacturing (total)	129.1	143.0	150.4	146.9	143.9
Food/kindred	7.6	8.6	8.4	7.4	6.7
Textiles	9.6	10.2	10.4	10.3	10.2
Paper/pulp	2.7	2.9	3.0	3.1	3.1
Printing/publishing	8.0	9.1	9.4	9.3	9.1
Chemicals	8.3	9.1	9.4	8.4	7.8
Petroleum refining	2.5	3.3	3.8	3.5	3.2
Other nondurables	4.0	5.6	6.3	6.0	5.9
Lumber/furniture	6.9	7.2	7.9	8.2	8.7
Stone, clay, and glass	4.6	5.4	5.0	4.3	3.8
Primary metals	1.4	3.3	3.7	3.3	2.8
Fabricated metals	14.4	14.1	14.9	15.1	15.4
Nonelectrical machinery	37.1	41.1	44.0	44.7	44.4
Electrical machinery	9.4	10.6	12.0	12.2	12.3
Transportation equipment	5.8	7.2	6.7	5.4	4.6
Other durables	5.4	5.3	5.5	5.7	5.9
Transportation, communication and public utilities	33.8	32.1	32.1	32.1	32.2
Trade	146.8	143.0	144.7	143.9	145.1
F.I.R.E.	70.8	73.0	77.4	77.9	76.0
Services	120.7	136.6	143.7	142.8	142.5
Total	642.0	691.2	720.9	709.9	701.8

Source: The Bureau of Business Research, Graduate School of Business, the University of Texas at Austin, 1983.

TABLE E.17
JOBS CREATED THROUGH INCREASED DEFENSE SPENDING IN TEXAS (results from input-output model)

	1983	1984	1985	1986	1987
Agriculture	596	618	800	665	441
Mining	3,439	3,454	4,399	3,585	2,385
Construction	776	801	1,060	889	594
Manufacturing (total)	18,586	19,852	26,468	22,511	14,108
Food/kindred	684	704	903	748	496
Textiles	35	35	46	38	25
Lumber/furniture	110	114	150	127	84
Paper/pulp	135	139	181	152	101
Printing/publishing	138	147	196	166	110
Chemicals	847	901	1,181	981	660
Petroleum refining	2,258	2,257	2,886	2,368	1,575
Other nondurables	117	126	169	143	96
Stone, clay, and glass	105	110	145	122	82
Primary metals	347	373	501	424	300
Fabricated metals	1,934	2,080	2,777	2,499	1,605
Nonelectrical machinery	1,160	1,275	1,732	1,473	1,000
Electrical machinery	3,040	3,316	4,488	3,802	2,566
Transportation equipment	6,100	6,624	9,050	7,640	5,073
Other durables	1,543	1,646	2,196	1,822	1,228
Transportation	2,098	2,149	2,787	2,296	1,529
Communication	732	769	1,019	959	581
Public utilities	1,180	1,224	1,602	1,331	892
Trade	3,661	3,907	5,135	4,303	2,883
F.I.R.E.	1,288	1,344	1,770	1,479	988
Services	6,005	6,250	8,216	6,847	4,586
Total	38,365	40,374	53,259	44,871	28,992

Source: IC^2 Institute, University of Texas at Austin, 1983.

TABLE E.18
U.S. NATIONAL DEFENSE RESEARCH, DEVELOPMENT, TEST, AND EVALUATION (RDTE) (in millions of dollars)

Budget Authority Major Missions and Programs	1974 Actual	1982 Actual	1983 Estimate	1984 Estimate	1985 Estimate	1986 Estimate
Research, development, test, and evaluation	8,582	20,060	22,805	29,622	32,206	34,145
Total DOD, military	77,550	213,751	239,407	273,400	321,600	356,400
RDTE (%)	11.0	9.4	9.5	10.8	10.0	9.6

Source: IC² Institute, the University of Texas at Austin, 1983. Based on budget FY 84 data.

TABLE E.19
U.S. FEDERAL GOVERNMENT ANNUAL INVESTMENT GROWTH IN RESEARCH AND DEVELOPMENT FOR SELECTED BUDGET FUNCTIONS OVER $1 BILLION (in billions of dollars)

Budget Functions	1974	1982	Proposed 1984
National defense (RDTE)	8.58	20.06	29.62
Space research and technology	2.96	5.54	6.52
Health	1.65	3.84	4.35
General science and basic research	1.01	1.54	1.94
Energy	0.52	2.72	2.07

Source: IC² Institute, the University of Texas at Austin, 1983. Based on budget FY 84 data.

TABLE E.20
U.S. GENERAL SCIENCE, BASIC RESEARCH, AND SPACE RESEARCH AND TECHNOLOGY (primarily physical science and engineering)

	In Millions of Dollars					
Major Programs Budget Authority	1974 Actual	1982 Actual	1983 Estimate	1984 Estimate	1985 Estimate	1986 Estimate
General science and basic research						
National Science Foundation (NSF)	$ —*	$1,006	$1,099	$1,297	$1,297	$1,297
Energy-related general science (DOE)	—	529	532	645	759	744
Subtotal	1,017	1,535	1,635	1,943	2,057	2,042
Space research and technology (NASA)						
Space flight	1,694	3,601	4,109	4,049	3,699	3,058
Space science application and technology	947	1,392	1,568	1,638	1,819	1,828
Supporting space activities	322	544	610	838	836	816
Subtotal	2,963	5,537	6,287	6,517	6,354	5,702
Totals	$3,980	$7,072	$7,922	$8,460	$8,411	$7,744

	In Percentage of Total Budget Authority					
Major Programs Budget Authority	1974 Actual	1982 Actual	1983 Estimate	1984 Estimate	1985 Estimate	1986 Estimate
General science and basic research						
National Science Foundation (NSF)	—	14.2	13.9	15.4	15.5	16.8
Energy-related general science (DOE)	—	7.5	6.7	7.6	9.0	9.6
Subtotal	25.6	21.7	20.6	23.0	24.5	26.4
Space research and technology (NASA)						
Space flight	42.6	50.9	51.9	47.9	44.0	39.5
Space science application and technology	23.8	19.7	19.8	19.3	21.6	23.6
Supporting space activities	8.0	7.7	7.7	9.8	9.9	10.5
Subtotal	74.4	78.3	79.4	77.0	75.5	73.6
Totals	100	100	100	100	100	100

*Data not availability.

Source: IC² Institute, the University of Texas at Austin, 1983. Based on budget FY 84 data.

Index

About the IC^2 Institute

The IC^2 Institute at the University of Texas at Austin is a national center for the study of innovation, creativity and capital. IC^2 studies are designed to develop alternatives for private-sector action aimed at regional and national goals.

Some of the specific areas of research and study concentration of IC^2 include the management of technology; creative and innovative management; measuring the state of society; dynamic business development and entrepreneurship; new methods of economic analysis; and the evaluation of attitudes, concerns, and opinions on key issues.

The institute maintains a strong interaction between scholarly developments and real-world issues by conducting a variety of conferences. IC^2 research is published in a series of monographs, policy papers, technical papers, research articles, and books.

About the RGK Foundation

The RGK Foundation was established in 1966 to provide support for medical and educational research. Major emphasis has been placed on the research of connective tissue diseases, particularly scleroderma. The foundation also supports workshops and conferences at educational institutions through which the role of business in American society is examined. Such conferences have been cosponsored with the IC^2 Institute at the University of Texas at Austin and the Keystone Center in Colorado.

The RGK Foundation Building has a research library and provides research space for scholars in residence. The building's extensive conference facilities have been used to conduct national and international conferences. Conferences at the RGK Foundation are designed not only to enhance information exchange on particular topics but also to maintain an interlinkage among business, academia, community, and government.

List of Contributors

The Honorable Lloyd Bentsen
U.S. Senator (Texas)

The Honorable Henry Cisneros
Mayor of San Antonio

Francis P. Cotter
Vice-President, Government Affairs
Westinghouse Electric Corporation

Brigadier General Robert J. Dacey
U.S. Army Corps of Engineers

Doctor Frank P. Davidson
Professor, Macroengineering Research Group
Massachusetts Institute of Technology

Doctor D. Keith Dodson
Vice-President, Brown and Root

John J. Egan
Manager, Business Planning Group
Coopers and Lybrand

Doctor Ken Gordon
Staff Director for Research, Development, and Manufacturing
President's Commission on Industrial Competitiveness

William Gregory
Editor-in-Chief
Aviation Week and Space Technology

Cordell W. Hull
Vice-President and Financial Officer
Bechtel Corporation

Admiral Bobby Ray Inman (Retired)
Chairman, Chief Executive Officer, and President
Microelectronics and Computer Technology Corporation

Joseph Kigen
Director, Government/Public Affairs
Westinghouse Electric Corporation

Doctor J. R. Kirkland
Kirkland Patton Associates

Doctor Eugene B. Konecci
The Kleberg-King Ranch Professor of Management
The University of Texas at Austin

Doctor George Kozmetsky
Director, IC^2 Institute
The University of Texas at Austin

Doctor Robert Lawrence Kuhn

Kathleen J. Murphy
Independent Consultant

Robert Ortner
Chief Economist
U.S. Department of Commerce

Doctor Raymond W. Smilor
Associate Director, IC^2 Institute
The University of Texas at Austin

Emil M. Sunley
Director of Tax Analysis
Deloit Haskins & Sells

Doctor James Wade
U.S. Department of Defense

The Honorable Mark White
Governor of Texas

About the Editors

EUGENE B. KONECCI

Dr. Konecci is currently the Kleberg-King Ranch Centennial Professor of Management, Graduate School of Business, University of Texas at Austin; Harry H. Ransom Centennial Fellowship, IC^2 Institute; and Professor, Department of Medicine, University of Texas Health Science Center at San Antonio. The Kleberg Professorship was established by the "King Ranch Family," and Dr. Konecci has held this position since January 1967. In 1983, it was endowed as the Kleberg-King Ranch Centennial Professorship in Management. Dr. Konecci joined the University of Texas in December 1966 and was appointed the first triple professor, i.e., Management in the Graduate School of Business, Professor of Aerospace Engineering, College of Engineering, and Professor of Bio-Engineering at the University of Texas Medical School in San Antonio.

ROBERT LAWRENCE KUHN

Dr. Robert Lawrence Kuhn is Senior Research Fellow in Creative and Innovative Management at the IC^2 Institute at the University of Texas at Austin and an Adjunct Professor of Corporate Strategy in the Department of Management and Organizational Behavior at the Graduate School of Business of New York University. He is also Adjunct Professor of Biotechnology and Public Policy at Hahnemann University in Philadelphia, where he works with the president of the university in strategic planning and the formation of new research institutes.